JUL
2006

W9-AJU-065

AFTER
DOLLY

W. W. NORTON & COMPANY

NEW YORK LONDON

AFTER DOLLY

THE USES AND MISUSES
OF HUMAN CLONING

IAN WILMUT
and Roger Highfield

For information about permission to reproduce selections from
this book, write to Permissions, W. W. Norton & Company, Inc.,
500 Fifth Avenue, New York, NY 10110

Manufacturing by The Haddon Craftsmen, Inc.
Book design by Judith Abbate
Production manager: Amanda Morrison

Library of Congress Cataloging-in-Publication Data

Wilmut, Ian.
After Dolly : the uses and misuses of human cloning /
Ian Wilmut and Roger Highfield.— 1st ed.
p. cm.
Includes bibliographical references and index.
ISBN-13: 978-0-393-06066-9 (hardcover)
ISBN-10: 0-393-06066-7 (hardcover)
1. Cloning, Organism—ethics. I. Highfield, Roger. II. Title.
QU 450 W744a 2006
(DNLM) 101265629   2006002030

W. W. Norton & Company, Inc.
500 Fifth Avenue, New York, N.Y. 10110
www.wwnorton.com

W. W. Norton & Company Ltd.
Castle House, 75/76 Wells Street, London W1T 3QT

1 2 3 4 5 6 7 8 9 0

To the tens of millions of people

who will one day benefit

from research on cloning, embryos,

and stem cells.

CONTENTS

AFTER
DOLLY

INTRODUCTION

FEW CELEBRITIES have so besotted the world's media as she did. Her arrival caused a sensation, triggering an orgy of speculation, gossip, and hype. She posed for *People* magazine, became a cover girl, and even caught the eye of Bill Clinton. Plays, cartoons, and operas were inspired by her story, advertisers traded on her image, and her name seemed to be on everyone's lips. Those closest to her said the attention rather went to her head, and she soon started putting on airs.

Nor was her career slowed by motherhood; even when she had six young ones to care for, she continued to be in the news. But, as with so many global superstars who live life in the fast lane, it was destined to go wrong. She developed a cough and began to go into decline. After a few weeks, she died tragically young of lung cancer, even though she had never smoked. Her passing marked the end of a great celebrity life. But this was no ordinary diva, for she was a sheep.

So much for the breathless media account of the life and times of Dolly the sheep. Her story is by now as much a part of the history books as of the annals of science. When she was born in the

Dolly with her surrogate mother soon after her birth. The ewe is a Scottish Blackface and obviously could not be the natural mother of Dolly. (Photograph courtesy of the Roslin Institute.)

Roslin Institute, near Edinburgh, on 5 July 1996, Dolly marked the beginning of a new era of biological control. With a large team, I was the first to reverse cellular time, the process by which embryo cells differentiate to become two hundred or so cell types in the body.

We had defied the biological understanding of the day: development runs only in one direction in nature: cells in tissues as different as brain, muscle, bone, and skin are all derived from one small cell, the fertilized egg. Until Dolly was born, it was thought that the mechanisms that picked the relevant DNA code for a cell to adopt the identity of skin, rather than muscle, brain, or whatever, were so complex and so rigidly fixed that it would not be possible to undo them. This deeply held conviction was overturned by Dolly. She was the first mammal to be cloned from an adult cell, a feat with numerous practical applications, many of which will raise profound moral and ethical issues.

The term "clone," from the Greek for "twig," denotes a group of identical entities. In the case of Dolly, she was (almost) genetically identical to a cell taken from a six-year-old sheep, the nucleus from which had been transplanted into an egg cell from a second sheep, and then inserted into the uterus of a third sheep, and then a fourth, to develop. It was because the process started with a mammary cell from an old ewe that she was called Dolly—our affectionate tribute to the buxom American singer Dolly Parton.

While we now take for granted that it is possible to clone adult animals, the birth of Dolly shocked those in the general public who dwelled on the implications of reproduction without the act of sex. The feat shocked many in the research community too. Scientists were apt to declare that this or that procedure would be "biologically impossible," but with Dolly that expression lost all meaning. Some pundits have even said that Dolly broke the laws of nature. But she revealed, rather than defied, those laws. She underscored how, in the twenty-first century and beyond, human ambition will be bound only by biology and society's sense of right and wrong.

When we created Dolly, we were not thinking about rooms full of clones, or creating hillock upon hillock of identical sheep to guarantee a good night's sleep. We were not thinking about helping lesbians to reproduce without the help of a sperm bank or multiplying movie stars. We were certainly not thinking of duplicating dictators.

The tortuous tale of Dolly does not fit the storyboard of a traditional Hollywood movie: my single-minded struggle, against all odds, to fulfill my dream; strange goings-on in a subterranean laboratory one dark and stormy night; or how a tight-knit group of hirsute underdogs toiling in an obscure Scottish lab snatched the cloning prize from arrogant, clean-shaven, lantern-jawed scientists working in a well-funded North American powerhouse of genetics.

True, my team did sport a fair amount of facial and long hair, but it was a special combination of factors that led to success, along with a spark of serendipity. The right people with the right skills and

Roslin Institute seen from the air. The building formerly occupied by PPL Therapeutics is in the foreground, the main institute is the biggest group of buildings, and Dryden Farm, where Dolly was born, is in the distance. (Photograph courtesy of the Roslin Institute.)

understanding came together at the right time in the right place—the Roslin's collection of pristine molecular biology laboratories, hay-strewn barns, and up-to-date facilities for surgery, whether on live-stock or on living cells.

Dolly's story is all the more remarkable for other reasons. The classic scientific tale of rival teams racing for a prize, as vividly depicted in Jim Watson's *The Double Helix*, did not apply in our case. We were mostly spurred on by pure curiosity, though we did have in mind practical applications in research and agriculture. As in any government laboratory, research was a grind because money was always in short supply. Even more so in those days, when the science establishment was subjected to punishing cutbacks: on the day Dolly was unveiled, the government withdrew its funding for my project. The company that helped us create her, PPL Therapeutics, was our ally but not our friend in the sense that we shared information only on a need-to-know basis.

The group was led by myself with Keith Campbell, a cell biologist ten years my junior, who had built up a profound understanding of what makes cells tick. It is still a surprise that we were so effective together because neither of us is well organized. His adventurous nature and knowledge of cells complemented my own scientific curiosity, my experience with cloning, and my strategic, step-by-step approach to achieving the goal of the genetic transformation of livestock. I conceived the project, but it was thanks to Keith's inspiration that it succeeded.

We were united by more than this, however. Both Keith and I had become disillusioned with life at the Roslin. As an embryologist, I had been forced to give up my research on prenatal mortality to focus on the genetic alteration of animals, work that, ironically, would eventually give me huge satisfaction and pave the way to Dolly. Keith, as the new member of the team with some sway over the science, found it frustrating selling his ideas to an established group. He felt resented and was unhappy—at least until our work together started to pay off.

Both of us had thought about leaving the Roslin. Like me, however, Keith had pragmatic reasons having to do with home and family for remaining in a stunning part of Scotland. There was a castle atop an almost isolated rock to the south-southeast of Roslin village We were surrounded by a range of serious hills, uninhabited in parts, rising to almost 2,000 feet and stretching some sixteen miles from the outskirts of Edinburgh to Lanarkshire. The landscape of uplands, lochs, and glens was just the place to follow the traditional Scottish pursuits of drinking whisky, walking, and staring into space. It was an ideal place to raise young children. I loved to jog in the Midlothian countryside and became reasonably adept at, and addicted to, the peculiar Scottish game of curling (if you are unfamiliar, think of bowls on ice with brooms).

Roslin the place, not the institute, held us tight during those difficult days. Keith and I would continue to drive to the laboratory together and have animated chats about the details of cloning. We were both excited by the science. We both hailed from the Midlands. But we were no Watson and Crick engaged in a race with rivals. We had different styles and outlooks. We did not socialize much.

Our shared aim was to develop methods for nuclear transfer in sheep that could be used to introduce precise genetic changes. Superficially, it may seem paradoxical that the wherewithal to create genetically identical offspring can also be used to introduce genetic changes into animals: the answer to the riddle is that you can use cloning to make a whole animal from the one cell where you have succeeded in carrying out successful genetic surgery among the billions of cells in which the surgery was botched or incomplete or had failed. That would offer huge advantages over the old-fashioned brute-force method in which we had to insert endless embryos into sheep without any clear idea of whether the embryos had been successfully modified.

We had just £20,000, which a Roslin committee had offered us to do some more experiments because of the good progress we were making with cloning. Of course, because of our previous experience we would use sheep. That meant we had to deal with their notorious obstetric problems (shepherds say that sheep spend their time dreaming up new ways to die). We were slave to their breeding cycles, which meant intense work during the winter, when sheep mate and conceive. The point was, however, sheep were astonishingly cheap. In those days, it was possible to buy one at the market for less than a bottle of mineral water in a posh hotel and for around one five-hundredth the cost of a cow. However, we were confident that because of their similar reproduction and embryology any method developed in sheep would also prove suitable for cattle. In effect, the sheep were small cheap cows.

It all seemed so simple. We would use special cells from embryos—stem cells—for cloning. These grow in the laboratory, but retain many of the characteristics of embryo cells. This was the reason why I was optimistic that this approach to making genetic changes in animals would work. But there was a problem. We could not multiply those sheep stem cells in the laboratory (we still cannot). However, in our failed attempts we did grow more mature cells that had, as scientists say, begun to differentiate.

Thanks to the insights of Keith Campbell into the mechanics of the cell cycle, we went on to make a remarkable find: it was possible to put these more differentiated cells into a special state (a resting state, called quiescence, which is unattainable by embryonic cells), and we discovered that we could clone them. Megan and Morag, Welsh Mountain sheep, were born as a result of this crucial insight. They were the first clones from differentiated cells. We had failed to grow stem cells, but Keith's new method of cloning had proved more powerful than we had imagined. Keith had always believed that cloning with adult cells would become possible, and with this new encouragement we set out to test the idea. The birth of Dolly confirmed the point emphatically.

As I waited for Dolly to be born during the summer of 1996 my mood would swing between elation and fear. Keith and I were confident of success, but, with such a complex process, involving so many steps and people, failure was always a strong possibility. And if we succeeded, the thrill of it all would be tempered by the thought of the fuss that would follow. I am a private person, and I knew my life would not be quiet anymore.

And succeed we did, but only just—in 1 of 277 attempts. At birth, Dolly weighed in at 6.6 kilograms, heavy for a Finn-Dorset but not surprisingly so. Perhaps in another season we would have had twenty Dollys. But it was much more likely that we would have failed. Despite the emphasis on objectivity and rationality, even in science it helps to be lucky. For Keith, however, Megan and Morag were the real stars of the scientific story. He considered Dolly merely "the gilt on the gingerbread." We had agreed that he would hold the coveted position of first author of the Megan and Morag paper, but that I would be first author of the paper that described Dolly's difficult passage into the world.

We kept Dolly's existence secret for the first six months, as the paper's details of how we had created her were scrutinized by other scientists before they could be published in an academic journal. Once the news of her birth leaked out, in February 1997, she made headlines, capturing the imagination of commentators, columnists, and opinion writers across the planet.

"*Researchers Astounded . . . Fiction Becomes True and Dreaded Possibilities Are Raised.*" That was the quaint way that one newspaper, the *New York Times*, greeted the news of Dolly's birth, but it was not the only one to sound the alarm. "*Cloning Discovery Has Unleashed a Wolf in Sheep's Clothing,*" "*Cloned Sheep in Nazi Storm,*" "*Dolly Opens Door for Life after Death,*" "*The Clone Rangers Need to Be Stopped,*" and "*Golly, Dolly! It's the Abolition of Man.*" The *Weekly World News* quoted a terrified researcher who described how Dolly killed a lamb, then ate her victim: "When you do something to anger her, she looks at you with those intense red eyes—eyes full of hate."

With the benefit of hindsight, this outburst looks entirely predictable and understandable (except perhaps the red-eyed cannibal). But I am still disappointed by the sensationalism that is so prevalent in today's journalism, especially in the headlines. Although I knew that some of the possibilities raised by this form of cloning are indeed worrisome, I thought Dolly's birth would be celebrated because many more applications of cloning in the farm, clinic, and hospital are wonderful. How naïve I was!

Members of the American media gasped that Dolly's birth "seemed to come out of nowhere" (she was not born in America, after all). Some scientists went further. One Nobel laureate wrote to the journals and accused us of making a silly mix-up that meant Dolly was not a real clone at all. Others thought Dolly was a "clone alone," a freak accident that could never be replicated. Most were convinced that Dolly was something special, however. There was talk of her creation being biology's equivalent of the splitting of the atom.

I am now used to the wild swings of the media pendulum—from the sensational prospect of cures to visions of apocalypse—and have often had my views misrepresented to make a better headline by supposedly responsible newspapers. But when Dolly was born, even I got caught up in the hype. During the initial brouhaha, I found myself telling one reporter that Dolly "might reasonably claim to be the most extraordinary creature ever born." This was an exaggeration. But not much of one. President Clinton of the United States reacted to the "startling" news of her birth by asking experts to review the implications; the Vatican and the European Commission swiftly followed. Within a few months came the first claims that human pregnancies were under way with cloned embryos; similar rumors have circulated ever since, sparked in part by the efforts of cults, madmen, and maverick scientists.

In pubs, in offices, and on the streets, people could not stop talking about the idea that human cloning might soon become a reality. Newspapers asked whether it would be possible to clone the dead. Some deluded souls saw Dolly as a harbinger of a form of genetic immortality; many more were fired up by the talk of a looming ethical and political nightmare. Inevitably, Frankenstein provided a cultural reference point for the discussion. *Time* magazine began its fourteen-page special report by casting me in the lead role: "One doesn't expect Dr. Frankenstein to show up in a wool sweater, baggy parka, soft British accent and the face of a bank clerk. But there in all banal benignity he was; Dr. Ian Wilmut."

The cover of the German magazine *Der Spiegel* depicted a regiment of Hitlers, the implication being that cloning people would somehow facilitate some kind of science-fiction tyranny, where individual identity was lost to a nameless, faceless, and oppressive mass of pseudo-humans. The cover stated, "Wissenschaft auf dem Weg zum geklonten Menschen. Der Sündenfall." (Science on the way to cloned people. The fall of man.)

In private, I had talked over the possibilities of human cloning two years earlier with Keith Campbell, as he drove me to work. That was the summer that two of our ewes had been pregnant with Megan and Morag, our first celebrity clones. Only in the month after the news of Dolly became known did I air my thoughts in public.

I held my breath as I prepared to address the crowded meeting in

Washington, D.C., of the Senate's Labor and Human Resources Committee, which was held on the eve of the first meeting of President Clinton's Bioethics Advisory Commission. I was anxious about the forthcoming grilling. As it turned out, I was indeed taken aback—but because the reception was anything but hostile.

One American scientist said I would win the Nobel Prize. A senator described me as "one of the true trailblazers in human history." Another declared I had "broken the biological equivalent of the sound barrier." A wire report concluded that "the world's best known shepherd" was treated by Capitol Hill lawmakers "like a media superstar." I did not feel like one, however, dressed in my gray suit and faded tie. Nor did I deserve to take so much credit for an effort that had depended on the hard work of so many others.

Human cloning in all its forms was, of course, the main subject of discussion. One Democratic senator likened efforts to stop cloning to the Catholic Church's attempts to silence Galileo. "Human cloning will take place. And it will take place in my lifetime. And I welcome it," he said.

But I did not. I told the hearing that to clone babies would be unethical and inhumane. I backed any efforts that could be made on an international basis to ban the practice. Equally, I stressed how important it was not to shut the door on life-saving applications. It was here that I presented the idea of deriving cells from cloned human embryos either to study disease or to treat it. As I will explain in these pages, I remain passionately opposed to cloning babies. I have coauthored a paper in a scientific journal with the unusually blunt headline "Don't Clone Humans!" But I am an equally passionate advocate of a restricted form of human cloning.

LIFE OF DOLLY

As ANY VISITOR to my office will tell you, Dolly inspired my friends and colleagues to give me endless ovine paraphernalia, sheep clocks, sheep mugs, and the like. She inspired a whole new generation of young people to study biology. We still get a steady stream of let-tres and e-mails asking for information. She inspired many crackpots, lunatics, and frauds too.

Meanwhile, at the focus of the furor the Finn-Dorset sheep seemed oblivious to all the angst she had caused. While hands were wrung and brows furrowed, Dolly lived a calm and cosseted life in a barn on the Roslin Institute site with two other sheep. These were Megan and Morag. Even though they had marked the first demon-stration that the developmental programming of a differentiated cell could be wiped clean so that life could start afresh, Dolly had pushed them both out of the limelight.

Unfortunately for her peers, Dolly behaved as if she were already aware of her celebrity status. She was greedy and would often grab more than her fair share of the "cake" made of barley and molasses that we fed them. She would also assert her authority by upending her trough as soon as she finished eating. This display of dominance would culminate with her placing her forefeet on the trough, puffing out her chest, and preening. She seemed to know that she was the alpha sheep of the flock.

I still think of her as the friendly face of science, however. Whereas most sheep are shy around humans, and huddle at the back of their pen when visitors arrive, Dolly seemed to bask in their atten-tion. She trotted to the front of her stall when she saw people com-ing by. The woolly prima donna would bleat excitedly and jump up—front feet on the railing—so that she could pose for the cameras. I should emphasize that this is odd ovine behavior: sheep do not nor-mally stand on their rear legs. She was an incredibly personable ani-mal and was interested and curious in the world around her, at least as much as a sheep can possibly be. Bill Ritchie, who selected the cell used to create her, remembered how, when he shouted her name, "you would get this deep bleat from the shed." I suspect her sweet and outgoing nature was shaped by Pavlovian training because she received more than her fair share of edible bribes to encourage her to flirt with photographers or appear at the front of her pen on cue for the legions of TV crews who came to visit her. In persuading Dolly to meet a visiting dignitary, or get her to move around, we would often offer her treats.

Visitors wanted to feed her too, but we urged them not to, explaining that she had a weight problem. In truth, we had contributed the most to her expanding girth. We provided the edible inducements, and she did not get much exercise. Eventually, Dolly joined the world's growing number of clinically obese individuals. We put her on a diet. Even her new food regime had a celebrity ring to it. "Hay only," read the sign hanging on her pen. Not quite the "Hay Diet" but close enough.

Like all her woolly peers, Dolly hardly looked as though she could change the course of history. But she did, of course. She met reporters, photographers, and film crews from across the planet. She inspired cloning references in *The Simpsons*. She was interviewed by Fay Weldon, author of *The Cloning of Joanna May*, who later said of the experience, "I thought she was such a beautiful creature, and so clearly had a soul, that I lost all fear of cloning." On Dolly's birthday, people would telephone to wish her many happy returns.

Dolly was fleeced in the name of medical research (her gray-white wool ultimately went to the Science Museum as a sweater designed by a schoolgirl in a competition run by the BBC's Clothes Show), and she posed with one Japanese couple for wedding photographs. In fact, she was photographed as much as a Hollywood movie star, more than any other sheep in history. As an experimental animal, one whose care was subject to strict guidelines, Dolly was banned from opening shopping malls or attending agricultural shows (despite innumerable requests) and had to turn down the invitation to mingle with fellows of the Royal Society, the world's oldest scientific academy, at their summer exhibition (a cardboard cutout went instead). The society, near Pall Mall, was a favorite haunt of scientific superstars, from Newton to Crick, but Dolly was not going to stroll through its marbled reception hall. For her, the journey to London was all risk and no benefit.

Even though Dolly was confined to the Roslin, she helped to introduce me to royalty. Thanks to Dolly, I have met the Queen twice, been presented with a medal by Princess Anne, and reprimanded by the Duke of York when my institute struck a deal with an American company, rather than a British one. In the Academy of Achievement held at Jackson Lake in the Rockies, an event that aims to inspire young people, I met Tom Clancy, that author of high-octane thrillers, and George Lucas of *Star Wars* fame and watched as the artist Dale Chihuly created his vibrant glass sculptures.

Also attending were the great and the good, from Colin Powell and Dick Cheney to computer wizards, astronauts, authors, and Nobel laureates. Over the years I have also talked to many politicians, from the former father of the House of Commons, Tam Dalyell, to Edward Kennedy (I remember his big, firm handshake). And, of course, I have encountered swarms of reporters, photographers, and anchormen and anchorwomen from around the world. All of this I owe to Dolly.

By the standards of her fellow sheep on the farm, many of which were slaughtered as early as nine months, Dolly lived a very long and full life. Some scientists feared, though for no clear reason, that she was sterile, but she defied her critics and went on to breed with a Welsh Mountain ram called David and have a daughter in April 1998.

Dolly with three of her lambs, all born naturally. (Photograph courtesy of the Roslin Institute.)

Her firstborn was named Bonnie because, as her vet Tim King commented, "She is a bonnie wee lamb." Dolly had two more lambs in 1999. In all, she gave birth to six lambs, all conceived naturally and all born healthy. She was the living, bleating, and woolly proof that new life—in the full sense of being able to reproduce—could come from a cloned adult cell.

Dolly became an unlikely icon for the promise and threat of biotechnology. She was featured in a "digital video opera" by the American composer Steve Reich, making its debut, as one headline put it, at the "*Baa-rbican.*" The ethical dimensions of cloning that she raised were explored by Caryl Churchill's thought-provoking play *A Number*. Appropriately enough, Scottish antimonarchists elected her as their preferred queen.

The company Zanussi used her image in an advertisement for washing machines with the slogan "Misappliance of science" (we applied for trademark protection of Dolly's image to prevent these depressingly downbeat campaigns). Many wanted to oblige. Richard Seed, an elderly Chicago physicist (not even a physiologist), in late 1997 announced his intention of cloning a human being within two years. Soon members of a religious cult based in Canada, the Raelians (followers of Claude Vorilhon, a French-born race car driver and journalist who says that he was given the name Rael by four-foot-high extraterrestrials), revealed a grandiose vision of achieving immortality by human cloning and founded the company Clonaid, run by Brigitte Boisselier, to carry out its mission. Followers believe that humankind started with the cloning of aliens 25,000 years ago, are advocates of free love (condoms mandatory), and believe that a superbeing called Elohim will return to Earth in 2025 to liberate believers.

Other would-be cloners could not be so easily dismissed. In March 2001, at a press conference in Rome, an Italian and an American announced plans to undertake human reproductive cloning. The Italian member of the team was Severino Antinori, a gynecologist notorious for having used donor eggs and *in vitro* fertilization (IVF) to make a sixty-two-year-old woman pregnant in 1994. The American was Panayiotis Zavos, a reproductive physiologist and an IVF expert at the Andrology Institute of America, in Lexington, Kentucky. Both Antinori and Zavos glossed over the technicalities and the many concerns about the effects of cloning.

Given the unusual way she came into being, Dolly's every bleat and every baa were by then being analyzed for their biological significance and the merest hint of decrepitude. She showed no real signs of being any different from the rest of the flock, however. But there were limits to what we could do to check she was really "normal." For example, we could not scan for aberrations of ovine mind or mood that could be linked with cloning. No one knows the mind of a normal sheep, after all.

After five years free of health problems came the first sign of trouble. Dolly began to limp, eventually becoming lame in her left hind leg. She was sent a couple of miles away, to Edinburgh University's veterinary school. X-rays confirmed that she had arthritis in her left knee. The condition was treated effectively by injections so that she had no discomfort. Nonetheless, my secretary was bombarded with calls from companies offering arthritis cures, pleading with us to try them out on Dolly.

Although arthritis is a fairly common ailment for a middle-aged sheep, it is less common to find it in the knee (the elbow is the joint that is usually affected). Perhaps, however, it was inevitable for a corpulent sheep who had been indulged all her life and liked to stand up and beg on her rear legs. But her limp did raise the first serious questions about the safety of cloning and whether it led to premature aging.

THE DEATH OF DOLLY

AFTER ENJOYING several years as the world's number one celebrity sheep, Dolly developed breathing problems and a cough in February 2003. We suspected that she was suffering from pulmonary adenomatosis, a disease that is not uncommon in adult sheep in which a progressive lung tumor causes emaciation, weight loss, and shortness of breath. There is no effective treatment, and, within a few months, infected sheep often succumb to the combined effects of the slowly growing tumor and pneumonia.

We were reasonably confident of the diagnosis because the disease had already struck four of the sixteen other sheep that had been in the same barn. The victims included one of Dolly's lambs and Morag (her barnmate and twin sister, Megan, was left unscathed, underlining that environment is as important as genetics). The disease is spread by infected droplets in the breath. Unfortunately, by the time the first sheep succumbed to the illness, it was too late to stop the spread of infection, since the disease takes up to three years to incubate. All we could do was ban any more sheep from entering the barn.

Sheep kept in indoor pens rather than open pasture are more prone to catch this kind of lung disease. Perhaps if Dolly had been allowed to roam the green fields, gulp in fresh air, and gaze at the Pentland hills, she would not have succumbed. But, because of her celebrity status, she had to be protected from lunatics who would want to harm her, criminals who wanted to kidnap her, and local students who might pull pranks.

To confirm our diagnosis, Tim King decided to send Dolly for a CT scan—where a cross section of the body is taken with X-rays—at the Scottish College of Agriculture (a highly unusual procedure for a sheep, but she was special, after all). That same afternoon, I was attending a meeting on the first floor of the Roslin Institute. The discussion ended when my secretary, Lynne Elvin, came in to tell me that Tim wanted to talk to me urgently. We went downstairs to my office, and I took his call. Dolly's infection was worse than we had thought. The CT scan had confirmed the diagnosis and revealed the full extent of her lung tumors. Dolly was also struggling to recover from the general anesthetic used to still her for the scan. Before her death, which now looked inevitable, Dolly faced weeks of pain and discomfort.

For a moment there, in my office, I held my head in my hands. It was a much bleaker picture than I had expected, and I was both taken aback and genuinely sad. We'd all grown fond of Dolly and her little quirks, and the prospect of losing her in such unfortunate circumstances was a jolt. Nevertheless, there seemed only one humane response. At around 3:30 p.m. on that day, Valentine's Day, I decided her suffering should end. My pioneering sheep was never allowed to recover from the effects of the general anesthetic. Tim King put her down with a lethal injection of barbiturates. "Dolly was almost a pet among the staff," King recalls. "There were many very upset people that afternoon. Tears were shed."

Years before her death, when she was only two, the Royal Museum of Scotland had approached us to make arrangements to preserve her in the event of her demise. Dolly was skinned by taxidermists from the museum on the very same afternoon she died to ensure there would not be any decay. Her skin was pickled and tanned to preserve it, then her hide was stretched over a fiberglass mold of her body, into which glass eyes were later inserted. Her nose and other muscles were fashioned of plasticine to give her shape, and

Dolly with Harry Griffin who as deputy director had responsibility for public relations and both managed and participated in this work. He is seen at the time when Dolly was unveiled in the Royal Museum in Edinburgh. (Photograph courtesy of Murdo MacLeod.)

her stuffed body was eventually put on display in the Royal Museum of Scotland, mounted on a straw-covered plinth.

Immediately after she was skinned, Dolly's remains were sent for a thorough postmortem examination, which was carried out by Susan Rhind, a senior lecturer in veterinary pathology at the University of Edinburgh who has done much to document the potential harmful effects of cloning, mostly on lambs. Susan and her assistant did a thorough job, making exhaustive tests and taking more tissue samples than usual, given the historic importance of their subject. Even though there was scientific interest in the viability of clones of clones, we had decided not to take samples of tissue for that purpose, and simply concentrated on assessing Dolly's physical state.

Susan confirmed that Dolly had pulmonary adenomatosis. Dolly's lungs contained large areas of firm, gray solid tumor. Around the central mass the cancer had spread to seed smaller nodules and masses into the surrounding lung tissue. There were also signs of pleurisy—her lungs were attached to the wall of her chest in places. Her larger airways and trachea contained white, frothy fluid. "It was a classic case," she said.

The disease is caused by a virus, called the jaagsiekte sheep retrovirus (also known as pulmonary adenomatosis virus). The virus infects lung cells responsible for secretion of lung fluid (alveolar type II cells); by transforming these cells and thereby expanding their numbers, it boosts lung fluid secretion and virus production. Cancer results because the envelope of the virus contains a protein that makes infected cells divide uncontrollably.

Susan also revealed the extent of the arthritis in Dolly's hind legs. Her knee—her stifle joint—in her left leg was the worst affected of all. Concern over potential harmful effects of cloning was now being voiced with increasing frequency, both in academic circles and in the media. Perhaps Dolly, being the first clone of an adult cell, would reveal all kinds of defects. But in fact the postmortem had revealed nothing particularly unusual for an animal of her age and her weight.

Because I was keen to end Dolly's suffering, I had not given any thought to dealing with the media. We initially thought we had better keep her sad passing to ourselves over the weekend so that we would have time to inform the acting director, Harry Griffin, who was away from the office, and prepare a full press release. But so many people had become involved with Dolly's last day in one way or another that the news was bound to get out.

We decided to make Dolly's death public immediately; any hint of secrecy would have fueled the fertile imagination of journalists who would have speculated on what it was about her death that made us want to cover it up. We did our best to make the facts known. Even so, her death was often (wrongly) assigned to premature aging caused by cloning. E-mails and cards followed, from scientists and public alike, saying how sorry they were that she had gone. The phone did not stop ringing for a week or so thereafter.

Dolly was survived by five of her lambs, and I am sure they remembered her. Work by Keith Kendrick at the Babraham Institute, Cambridge, suggests that sheep have a much richer inner social and emotional life than we have given them credit for. They can remember at least fifty sheep faces for many years, even in profile, when most humans would be pushed to tell the animals apart. These animals can also remember ten or more familiar human faces, and their brains, like ours, seem capable of forming mental images.

So, while apparently ruminating mindlessly, Dolly could be dwelling on long-absent flockmates, mothers, or even a significant human in her life, like me. Even in her woozy state after her anesthetic, Dolly may have been able to detect the discomfort of Tim King when he ended her life. Sheep, like a number of other animals, are better than people at detecting bodily changes occurring during emotional or psychological states in others, even if they cannot imagine what that other creature is actually thinking.

The Monday after her death, on 17 February, I held a "wake" at the Roslin to note Dolly's passing, to signify the importance of the experiment, and to say thanks to all the people who looked after her. With champagne we toasted Dolly to celebrate her extraordinary life. The carers, vets, and scientists had a chance to say thank you to this remarkable animal. Her passing was a sad time for those, like me, who had seen her from when she began her life in a Petri dish, visited her often in her stall, shown her off to the world, and discussed her almost every day of her life. When I go near the barn where Dolly lived, I am reminded of her.

Many media commentators concluded that cloning was to blame for her death. Typically, sheep can live eleven or twelve years, yet Dolly managed only six. One newspaper announced, "Cloning may have consequences far beyond a creature's birth." Another stated, with grim relish, "Dolly the sheep may have been fated to die prematurely from the moment she was conceived." She did indeed seem to have inherited six years of wear and tear from the udder cell used to clone her: studies by Wei Cui, one of my colleagues, found evidence from structures in her cells called telomeres that Dolly's biology was older than her years—her telomeres were shorter than expected and more typical of an older sheep. Was it that, like most copies, she had faded a little bit quicker than the original?

Perhaps. But you can't infer much from a single case. Particularly when there were no sheep to compare her with: we don't know enough about the typical illnesses that affect a six-year-old sheep. In fact, no one could even state the "normal" expectations of aging or her average life expectancy with confidence, since we tend to eat most sheep by the time they are about nine months old. From the example of Megan, one of the clones in Dolly's shadow, Dolly did seem to die young; I celebrated Megan's tenth birthday, on 19 June 2005, with members of the original team who helped create her.

But Dolly's progressive lung disease, which had also claimed the life of Megan's twin, Morag, probably said more about being kept in a barn than about being cloned. To confuse matters more, there is also individual variation in any animal species. Other studies, to which I will return, say much more about the dangers of cloning, from supersized offspring and breathing difficulties to immune system problems and blighted pregnancies. Dolly's postmortem revealed a specimen that was unremarkable for a middle-aged sheep with a weight problem that had enjoyed a celebrity lifestyle.

There was nothing exotic about the viral disease that prompted her euthanasia. Dolly was unscathed by cloning, even though she was a pioneer of the technique.

AFTER DOLLY

NEAR THE Roslin's Large Animal Unit, a few yards from a public path where thousands of people stroll each year, is an unremarkable building that is still called the Visitors' Centre. Rather than containing posters, coffee machines, and souvenirs, the center is actually a barn, adjacent to a small field, and was so named because of the hundreds of journalists, politicians, dignitaries, and others who had made the pilgrimage there to meet Dolly. Years after her death the name stuck, a tribute to the star qualities of this animal.

Throw open the doors today, and you will confront a little flock of five clones in a straw-lined pen, the only living evidence of the years of work by our large team: Megan; two sheep from cloning experiments in which an attempt was made to implant a human blood-clotting factor; Dolly's daughter Bonnie; and Morag's offspring, Katy. Since those heady, pioneering days, the focus of the effort to clone farm animals has moved abroad. The team that created Dolly has split up. Keith Campbell moved to PPL and then to Nottingham, where he is trying to improve the process of cloning. PPL is, in effect, no more. And I now work for Edinburgh University. The Roslin, with which I still collaborate, is focusing its efforts on farm animal breeding and health.

The life of Dolly opened up thrilling new opportunities in medicine and, by the same token, raised serious ethical issues and stirred many concerns. And continues to do so. Even as I write this, I am feeling the aftershocks of Dolly's birth. I have in the space of a few days been criticized by an American presidential ethics adviser for being "misguided." As he declared that genetically engineered humans would never happen, he raised the ludicrous possibility of designing people with wings (was he thinking of angels, perhaps?). I have also been picketed by protesters in sheep masks and quoted in a tabloid article headlined, "*I'm Not God.*"

I believe in the right to protest. I believe equally that right-wing religious paranoia is slowing the quest for treatments and, as a result, will harm people and cause suffering. I find it a constant cause of frustration that, in all the public debate, the harm that can be done to future generations by neglecting a useful technology is rarely taken into account. Disregarding the hype, of which there is much, I have no doubt about the long-term potential. The creation of cloned embryos from patients will, over the next few decades, lead to treatments for degenerative diseases such as heart disease, spinal cord injury, liver damage, diabetes, Parkinson's, motor neuron disease, and Alzheimer's—all of which cause damage to cells that, subsequently, are not repaired or replaced. By creating a cloned embryo of a patient, we can obtain a source of the patient's own cells—stem cells—that can be used to understand the disease, test treatments, and not only repair a body but regenerate it too. They can metamorphose into any cell type.

We cannot know yet which of these serious diseases will be treated with cells from cloned embryos or when treatment will begin, but I believe that the drive to perfect them is strong because there is no fully effective treatment for any of them and in some cases there is none at all. One day doctors will be able to use cloning to grow a patient's own cells and tissues to carry out repairs. Cells from these embryos will also speed the search for the next generation of block-buster medicines and help reduce our dependence on animal research. As a bonus, this work will give profound insights into human development and how it can go wrong and into how to correct many terrible genetic diseases in the embryo.

The potential of cloning to alleviate suffering—even end it for some diseases—is so great in the medium term that I believe it would be immoral not to clone human embryos for treatments. In the long term, a vast range of alternative and embryo-free ways to grow cells and tissues, perhaps even organs, may also rest on the foundations of this research. Cloning may by then have been a mere diversion that will eventually be superfluous.

However, I want to go even further than this and propose that scientists may one day grow cloned human embryos to term to prevent the suffering caused by hereditary disease. Doctors should be able to offer at-risk couples the opportunity to conceive with IVF methods, break down the resulting embryos into cells, correct any serious genetic defects in these cells, and then clone demonstrably healthy cells to create a new embryo that can be implanted to start a pregnancy.

This does add up to a qualified proposal to clone and genetically alter human beings, but I hope to build a case to show that this approach could eventually be justified because it does not, paradoxically, entail cloning a person (by the understanding of most people) and because it could do much to combat serious genetic disease and to reduce human suffering.

My vision raises many technical questions. Scientists still have much to do to understand the process of cloning and the detailed molecular mechanisms of how it goes wrong and why it does so often. Only then will they find ways to make the process of cloning safe. Only then will they have complete confidence in growing a patient's cells, by the creation of embryo clones. Only then can they consider my proposal that for certain diseases we carry out the cloning of IVF embryos.

My vision faces stiff opposition. Many people will fight *any* proposal to create embryos. Many will object to *any* use of embryos in research. Many will object to *any* proposal to alter the genes of human beings. Many more besides are likely to feel profound unease at what I suggest. But I feel it is best to say what I think, rather than what people would like to hear, even though I suspect my peers will consider me at best brave, at worst foolhardy and naïve. Until now, it has been difficult to discuss these ideas in public. The drizzle of outlandish remarks and claims from the mavericks who say they are about to clone babies has distorted the media debate and fueled political efforts to ban all forms of cloning, both reproductive and therapeutic. In short, it has threatened to derail serious science.

One unsung and unlikely hero in the efforts to disentangle the complex issues so that my proposal, among many others by serious scientists, can be discussed in a clearheaded way is Bernard Siegel. Once the owner of Florida Championship Wrestling and of Miami Tropics, a minor league basketball team, Siegel entered the cloning story as a Florida attorney who had worked in family law. He turned his hand from everyday cases in Coral Gables to debating on television with Rael, the supreme leader of the Raelians, believers in aliens, and supporters of human cloning. Siegel sued Clonaid, an organization that claimed to have cloned a human—"Eve"—in December 2002, to have a guardian appointed for the alleged child. He reasoned that anyone reckless enough to create Eve was hardly going to make an ideal parent. At the time of writing, Clonaid had not come up with any evidence to convince scientists that Eve was cloned.

Siegel, dubbed "the clone ranger," moved on to educating United Nations delegates on the ins and outs of cloning, reproductive science, and stem cells. I was among the scientists who joined the advisory board of his nonprofit organization originally called the Human Cloning Policy Institute (now the Genetics Policy Institute) to try to raise the level of the debate and, in particular, to distinguish clearly reproductive cloning from the other uses, such as therapeutic cloning and the combining of cloning with genetic modification. We are still trying to make sure that life-saving and life-enhancing research is not ended along with attempts to clone babies.

At the heart of the discomfort with my proposal is the status of the human embryo and whether it counts as a full version of human life or lies somewhere between the status of a cell and that of a person. Even if one accepts this gray scale of life (and many people do not), does the embryo still deserve respect? And if it deserves respect, would that prohibit its use in research? Do the ends—new treatments for horrific diseases—always justify the means? There are broader questions raised by my agenda. What do we mean when we condemn research as unnatural? What do we actually mean by a human life and a person? What is consciousness? Indeed, what is life? In the following pages, I hope to shed a little light on the forces that are now shaping our destiny.

Lynne Elvin, personal assistant to Ian Wilmut since 2000, in the Roslin office shortly before they moved to the University of Edinburgh. (Photograph courtesy of the Roslin Institute.)

The two authors on the grounds of the Roslin Institute during the preparation of this book. (Photograph courtesy of James Fraser.)

ONE

CLONING
THE
CLONER

THREE JUSTIFICATIONS for human cloning tend to crop up more than any others in public discussions: the first is to overcome infertility, the second is to "bring back" a dead child, and the third is to duplicate someone of astonishing ability or talent—to give the world another Mozart, Curie, or Einstein.

But how about cloning a person who is far from being an ideal male specimen (even to his wife and children), a bald chap with a beard who notched up a somewhat undistinguished academic record at school? Someone who is famous for being associated with celebrity sheep? An individual who proves that scientists who do supposedly extraordinary things are far from being extraordinary, let alone a crazed Frankenstein or a Strangelove figure? An ordinary guy, like me. After all, if you prick me, do I not bleed? If you tickle me, do I not laugh? Of course I do.

I have been asked many times whether I would ever consider the ultimate act of vanity—cloning myself—in the years since my colleagues and I unveiled Dolly. There is, after all, a long and honorable history of medical scientists who have tested their ideas on the most curious of experimental subjects—themselves. Since the earliest days of medicine, researchers have swallowed microbes, injected vaccines, followed starvation diets, and even performed their own surgeries. Usually, researchers double as subjects to make their experiments both practical and ethical. But not in the case of cloning. Each time I have been asked whether I would use myself as a guinea pig, I have had to explain how, even if the many technical obstacles could be overcome and a mini me, "a little Ian Wilmut," could be born unharmed, the answer would be the same. No. I have no intention of cloning myself. One Ian Wilmut is quite enough.

Living in my shadow would be intolerable. Even though the reality is that a clone would be an individual living his own life, he would have to deal with the expectations of parents, family, friends, teachers, and, of course, the media. That would be a heavy burden. Imagine what a broody teenager would make of being told he is a genetic copy of a parent, let alone a cloner. Imagine what it would be like to know that you are the product of a scientific experiment. Imagine his irritation to learn from his pals that his father—also his brother—was called Wilf at school and that this would perhaps make a good nickname for him too.

Imagine living with the oppressive feeling that, both medically and psychologically, the future would be no longer open. Instead, he would fear that his lifetime could follow the same trajectory. My little identical twin would be an individual, a person in his own right, but would have to bear a heavy burden of growing up to feel like my duplicate. He would watch each serious illness that struck me with some trepidation, wondering whether he would suffer in the same way.

When a son or daughter lives in the shadow of a famous or high-achieving parent, there are often tensions. The unusual similarity and the unusual reasons for the birth of my clone, let alone my role in it and the controversy and furor that reproductive cloning has stirred, would inevitably create even greater and more unpredictable stresses. It seems inevitable that parents, relatives, and friends, although misguided, would expect the clone to be an echo of the original person (I have been asked again and again whether my children are biologists—for the record, Naomi is, but Helen is a mathematician and Dean, whom we adopted, works in hotels). This belief persists even though we know that each person is an individual and different from everyone else. There is still a strong expectation that we will be like our genetic forebears; this expectation would be greater, even overpowering, for an unfortunate clone. Reproductive cloning runs counter to our present culture and would wreak havoc with family law. One survey of more than 1,800 people for the UK Department of Trade and Industry, published in 2005, found that around two-thirds felt cloning was a "bad thing." Similar consultations have produced an overwhelmingly negative response: people don't want it.

CURSE OF THE CLONE

THE PSYCHOLOGICAL and social impacts of cloning have been explored by the psychiatrist Stephen Levick in his book *Clone Being*. This exercise has many different facets, because cloned children would have a confusing array of unusual relationships to relatives and would suffer what he calls "genealogical bewilderment." The clone would be the identical twin of the donor, who could also be his parent. In a genetic sense, the parents of the person copied would also be the "parents" of the clone. By contrast, the birth mother of the clone might make no genetic contribution if, for example, the child were a clone of her partner.

Lacking cloned children to study, Levick searched for relevant circumstances in modern life. He assessed situations that might be expected to reveal some of the effects of being a clone, such as that of identical twins, who are natural clones of each other, and children born through the use of *in vitro* fertilization (IVF) and other reproductive technologies. The parents of these children, if they are candid about their high-tech method of conception, may constantly remind their sons and daughters that they are "special." Louise Brown, the world's first IVF child, used to wonder whether she would grow up normal: "Am I going to grow up with two heads or something?" Schoolchildren would ask her how she fitted into a test tube. "I knew that it wasn't a test tube, and I used to get fed up of explaining it then. They couldn't understand that I was actually born from Mum."

Other important insights into the lot of the clone come from stepchildren. A stepchild is genetically related to only one parent, as would be the case in cloning where there is one genetic "parent." Here the genetically programmed rule—help your closely related kin—may have an ugly mirror-image, according to Stephen Emlen of Cornell University. Stepchildren, on average, suffer higher rates of physical and sexual abuse and even death than children in intact families. Stepparents tend to invest less time and effort in the offspring from their partner's previous marriage than they do with their own children. The incidence of sexual abuse of stepdaughters in one study was eight times that of biological daughters. Children in stepfamilies tend to leave home significantly earlier than children in intact families. And stepfamilies are, on average, less stable than intact families. The incidence of subsequent divorce is higher in second marriages and increases with the number of stepchildren present. Levick expresses concern that similar risks might face a clone. The genetic parent of the clone may feel the child is "mine" rather than "ours."

Levick says that any parent motivated to clone himself or herself is bound to feel let down by the results. The narcissist is always disappointed in his or her children and their failure to live up to "perfection." He would surely feel entitled to expect a clone to be just like him, or his ideal. Many people have already had to struggle under the burden of unrealistic parental expectations. For a cloned child, the pressure could be extreme, even though parents know firsthand that nurture is as important as nature.

"A clone would be likely to face extra difficulties in negotiating every stage of psychological and psychosocial development," Levick concludes. Despite his concerns about the likely psychological ill effects of being a cloned child, he does not recommend prohibition, however. "Cloning per se would not be wrong." A ban, such as the one already in place in the UK, would stigmatize any cloned children. They would be made to feel a deep sense of shame even though they would carry none of the blame. And a ban would force cloning underground, making it even harder to help any clones.

In an attempt to undo their terrible loss, some parents might want to clone their dead child from one of the infant's cells. Levick tried to anticipate the possible results by drawing on the example of bereaved parents who try to deal with their loss by having another child in the hope that the new child will be a replacement for the one who died. A clone would regard herself as a replacement, as a living memorial to the dead. She would have difficulty in developing her own unique identity. And she would be unlikely to fill the shoes of the ghost. The tragic results Levick forsees for such a child clone leads him to view any attempt to try to "replace" the dead by cloning as unethical.

There are also good scientific reasons why clones will never be copies. Genes are not as powerful as many people think. A clone may have identical genes to those of the lost child, but that is not enough to ensure he develops the same way. The influence of genes is modified significantly by that of the environment. Genes are in constant dialogue with their surroundings. They are in dialogue with the rest of the cell in which they reside, which is in dialogue with other cells in the body, which in turn is in dialogue with the world at large, through education and experience. This nested dialogue shapes the development of a fertilized egg into brains or brawn. Just as one can never really relive a moment, this dialogue can never be exactly reproduced. As I explain below, the personality of a clone will be individual. Cloning cannot resurrect a loved one for this reason alone.

DROP THE DOPPELGÄNGER

MANY OF THE psychological burdens of cloning arise because many people still think—wrongly—that cloning offers a way to copy another human being, just as a photocopier duplicates a document. They imagine that the process of cloning would seek to capture not just a person but a person at a particular moment in time, perhaps even to go back to a lost moment, to try to defy what the novelist Milan Kundera called the unbearable lightness of being. This cannot be done: even identical twins who have developed in the same womb and experience the same upbringing are never really identical. Genetic identity is not the same as personal identity, and selves, unlike cells, cannot be cloned.

Because of the influence of environment, luck, and circumstance, it is impossible that my clone could grow up the same way as me. He would grow up in a different time, in a different place, eating different food, hearing a language decorated with different slang and idiom. He would have different best friends, different teachers, different lovers, and so on. Take two people, one of whom spent her formative years in the postwar austerity of the 1940s and the other in the relatively well-off and liberal years of the twenty-first century. They could not possibly think alike.

Parents of natural clones—identical twins who have come from the same human egg—often dress them differently and emphasize differences in character in a bid to make them true individuals. This may be helpful for the twins, and their friends and family, but is unnecessary. From the very start of life, each embryo would have implanted in different places in the uterus, or drawn on different parts of the same placenta. As a consequence, they would have different access to food as well as different exposures to toxins and infections. They may start off with the same genetic code, but this can suffer different errors and mistakes as genes multiply in their bodies. The way genes are turned on and off by a process called imprinting can be patchy and vary between twins. Slight differences in the frequency of cell division or the rate of cell migration may slightly alter the appearance of one twin relative to the other.

The brain would be affected in a similar way. Small changes in the way the nervous system wires up and matures might also influence ability or behavior, although many other factors will be important as well in determining these characteristics. Identical twins are more similar than you and I are. But they are not really identical at all.

Plenty of evidence from animal cloning, from cats to mules, shows how environment shapes an individual as much as genes do. We cloned a quartet of young Dorset rams, Cedric, Cyril, Cecil, and Tuppence, from embryo cells that had been grown (cultured) in the laboratory. They were genetically identical to the same culture of Poll Dorset embryo cells and yet were very different in size and temperament. Although they were all more aggressive than females, as one would expect, Tuppence was by far the quietest of the four. Small variations in cell division and migration result in differences in the pattern of black-and-white skin in Holstein dairy cattle. The world's first cloned horse—Prometa, unveiled in 2003 by Cesare Galli of the Laboratory of Reproductive Technologies, Cremona, Italy—could be distinguished from her sister/mother (unusually, she was the clone of the horse that carried her) by a white stripe. And so on.

As variation is to be expected in any species, pet owners should not expect clones of Tiddles to be identical to their late lamented cat. This was observed with the first feline clones, created at Texas A&M by Mark Westhusin. Rainbow the cat was a typical calico with splotches of brown, tan, and gold on white. Cc (carbon copy), her clone, had a stripy gray coat over white. And while Rainbow was chunky, Cc was sleek; Rainbow was reserved, Cc was playful; and so on. The prospects of making exact copies of Patch the cat and Spot the dog look bleak.

HARD ACT TO FOLLOW

THE EXPERIENCES of twins and animal clones show that, because of tiny differences in the way genes are used in the body, the future of a clone is very much open. My own curriculum vitae illustrates the equally profound effects of chance, happenstance, and serendipity: the basic genetic clay of life can be pounded and shaped and molded into endless variations.

The Wilmut family consisted of modest, middle-class professionals. My parents both taught and, in those days, could afford to buy their own home without the need for special loans from the government. We enjoyed weekends and vacations in a trailer, touring Britain. My parents were especially fond of exploring the remoter reaches of Scotland where relatively unused single-track roads had grass growing in the middle.

Although I was born in Hampton Lucy, near Warwick, I was brought up in and around Coventry, where my father taught. Before World War II, it had been one of the finest preserved medieval cities in Europe. But when I ran around its streets as a boy it bore many scars. It had been at the industrial heart of Britain's war production engine, making it a major target of Nazi bombing raids. Unlike the village where I now live, it offered no panoramas, vistas, or horizons. Feeling claustrophobic, I would come to crave wide-open spaces.

The family moved to Shipley, in the West Riding, when I was six, and I attended the school where my father taught in Saltaire, the model industrial village built by the Victorian philanthropist Sir Titus Salt. While visting my grandfather in the coastal resort of Morecambe at the age of ten, I met an inspirational figure, a merchant navy captain. My future was by now quite obvious: I wanted to leave Shipley and sail the seas. Over the next few years, my interest in the ocean grew into an obsession. But there was one problem: a mutation of a gene on my X chromosome.

I am among the 7 percent of the male population who cannot distinguish red from green. I lacked the right stuff for the navy. When it comes to teenagers, that little word "no" is a poor persuader. Rather than criticize my chosen career path, my family adopted a more subtle way to convince the fourteen-year-old me that life on the high seas was unsuitable. An uncle who happened to be an optician sent me to the docks so that I could learn firsthand what would happen if I could not distinguish navigation lights, where port is red and starboard green. I quickly realized that Captain Wilmut would be a danger to himself, his crew, and other ships. Today my clone would find that the navy is happy to accept the color-blind, but only in certain careers. My clone could end up sailing the seven seas rather than a prisoner of the international conference circuit who flies across the world's waters and never sails upon them. If my clone knew much about my upbringing, he would be apprehensive about going to school. By the time I faced my first qualification exams, I had moved to Scarborough and, at a critical time when it came to my education, had been off school with appendicitis. I managed to get only four O levels at the age of sixteen and had to retake them. When the next academic milestone loomed—A levels—I was told that I was falling behind in chemistry. I made up for lost time and passed chemistry and zoology. But I failed botany.

I would love to tell some homespun story about how, despite this hammerblow to my ego, I had picked myself up, stuck out my jaw defiantly, and told my school friends that one day I would do something that would give the president of the United States pause for thought. But I had no delusions of grandeur and felt an average sort of schoolboy. My botanical setback merely underlined this fact. Perhaps, however, this low-key start to my career would take the pressure off my clone to perform.

Would my clone be lucky, like me, and fall into the arms of the love of his life at school? If not, he would lack what I consider to have been a key ingredient of my life, my emotional motor that has helped keep me going. I first met Vivienne when we were both studying for our A levels in Scarborough. She passed her A levels in mathematics, physics, and chemistry. She still motivates me today.

I feel I know her better than anyone else. Yet it is hard for me to imagine what would go through her mind if she had borne my clone and found herself looking after a schoolboy who looked just like the one she had fallen in love with decades earlier. How would Vivenne regard him? Some observers, like Levick, have even speculated that cloning would raise the specter of incest.

Four decades ago I found myself on farms and then attending agricultural college at Nottingham University, after my undistinguished performance at school. Farming was my next love, not least because I wanted to breathe fresh air. But I discovered that you need to have a good head for business too, and, in this important respect, I was lacking. At Nottingham I became gripped by a subject that has fascinated people since the days of the ancient Greeks—embryology, the study of how a minute speck turns into a living, breathing animal—and studied with G. Eric Lamming, an authority in the field of animal reproduction with a down-to-earth nature, a farmer at heart, who established Sutton Bonington as one of the top animal science research institutions in the world, where Keith Campbell would end up as a professor.

I enjoyed working in the laboratory and, in my middle year, decided to write to nearby research centers to find a summer job. A ripple of increasingly plaintive pleas for work spread out from my digs in Sutton Bonington to labs at progressively greater distances from Cheltenham, the Regency spa town in Gloucestershire where Vivenne was working and where I wanted to be. This led to a lucky break that would change my life. I won a scholarship to work at the Animal Research Station, Cambridge, which made major contributions to the development of methods for semen freezing and artificial insemination, embryo recovery, culture, and transfer in cattle. That summer Chris Polge, one of the great pioneers of animal reproductive science, both as an entrepreneur and as a researcher, took me on for a month or two. This marked the real start of my career.

FROSTIE

CHRIS IS A farsighted scientist, a patient teacher, and still a generous friend. During that first summer I worked for him, the station was host to Professor B. N. "Billy" Day from the University of Missouri-Columbia. He was visiting Britain to work on the estrus cycle of the pig, and how to control it, and a fine steak and rib barbeque was thrown in his honor by Chris and his wife, Olive, at their home in Girton. At this time, I had my first experience of wealth (at least what I thought of as wealth) and advanced office technology of the day: I caught my first glimpse of a £20 note, which I had been given to pay for the use of a photocopying machine, creating the quite misleading impression that scientific research was a way to make money (certainly not working for a British government lab, alas). I spent hours on the photocopier, copying experiment records, so that Chris and his American visitor could collaborate.

While I had to do my fair share of routine laboratory work, including looking after the pigs and cleaning up in the lab, I had my first chance to work with sperm and embryos, and I studied living embryos for the first time, watching them begin to organize themselves into a bag of skin with its associated bones, muscle, and plumbing, the fetus, and its ancillary equipment: the placenta, umbilical cord, and so on. I also learned the signs that would reveal which eggs were fertilized and developing normally.

Like many other students, I learned as much in the coffee break as in the laboratory. Sometimes the chat was about cricket, a subject close to the heart of Chris Polge. At other times we discussed gossip and personalities—usually of the scientific variety. Most of the time we discussed reproduction. Just exactly how does a mated sow know whether she is pregnant or not? It turns out that a pig carries a pregnancy only if she has four or more fetuses developing. If there are fewer, she comes back into heat to have a chance of producing a larger litter. I was hooked by these sorts of questions. I knew then that, if I could, I wanted to build a future in research.

After the summer work I found myself back at university, searching for a research post. I wrote to Chris later that year asking whether he would give me a reference. I did not ask for a job, because I knew there were none to be had. Here again, an effort to create a "mini me" would find it hard to re-create what happened next: few people have a career trajectory that goes in a straight line. But mine did at this crucial point.

Chris surprised me by asking me back, offering me a postgraduate position, funded with a scholarship from the Pig Industry Development Authority. It turned out that I had been lucky because of another's awful misfortune: one of Chris's team had committed suicide. My feelings were, of course, mixed. I was uneasy about the terrible circumstances, but this turned out to be an amazing break. Without it my career may well have fizzled there and then.

After finishing my university studies and achieving a good second-class honors degree, I returned to the Animal Research Station, where I found myself studying for my doctoral thesis at Darwin College, Cambridge, working on the freezing of boar semen. Most important, I was again under the spell of Chris, who, among his many achievements, had done the pioneering work in the fifties to show how a small molecule—glycerol—could protect frozen cells. He also had a management style that suited me: he was always available to challenge my ideas and offer advice but left me to get on with my research.

That first project would set a pattern for my subsequent work. Finding out how to succeed in freezing semen requires a systematic approach that would come in handy for perfecting cloning too. To freeze sperm successfully—so that they can be revived to wriggle once again—depends on the type of protective chemicals that are used, the rate at which the sperm are cooled, and many other factors. So many, in fact, that it is easy to get lost in a forest of permutations.

I had the good fortune to be taught by a Hungarian visitor to Chris's lab, Steve Salamon, who gave me many insights into the design of effective experiments that could provide a clear path through this jungle of variables without descending into the chaos of pure trial and error. Inspired by Salamon and with equipment designed by John Farrant in the Medical Research Council in Harrow, I used an apparatus to test several different freezing regimes, which I called my "four-seater loo." Each loo was a flask of liquid nitrogen in which embryos were cooled at rates varying from a fraction of a degree a minute to one hundred degrees Celsius a minute. It was probably around this time that I first became aware of cloning, notably the efforts to multiply rabbits in Oxford by Derek Bromhall, who would become one of the first victims of human cloning (more of him in the next chapter).

After I finished my Cambridge University doctorate in 1971, I was lucky to be able to remain working with Chris, this time funded by a fellowship from the Milk Marketing Board. Now defunct, this organization supported all aspects of milk production and distribution, including the breeding of superior cows. Research on cattle by biologists like Chris and geneticists based in Edinburgh had given Britain some of the finest (in terms of being productive) dairy cattle in the world. I was able to work with Tim Rowson, the vet at the station, from whom I learned the techniques of surgery, a skill that would one day come in useful for cloning.

We were the first to freeze a calf embryo successfully, thaw it again, and transfer it to a surrogate mother. The challenge was to stop ice crystals from forming, which would fatally damage the embryo. During cooling, as ice forms around the cells of an embryo, the salts in the remaining solution become more and more concentrated. As a result, water will flow out of the embryo into this solution, by a process called osmosis. I found the optimum cooling rate to stop ice crystals from forming in the embryo. Thawing must also be controlled carefully. Because it takes much longer than sperm to obtain the embryos and test their viability, this was a laborious project. But it was worth the effort.

The result was the world's first "frozen calf," a red-and-white Hereford Friesian cross. We decided to call him Frostie. The efforts that led to his birth in 1973 were filmed by a Dutch TV crew, who camped out in a barn for several days to be sure to witness the event. When Frostie was delivered by Tim Rowson, we celebrated with champagne, a suitably extravagant display prompted by the presence of cameras. I also got my first experience of dealing with the media. As the *Daily Mail* put it, "Ice-Age calf weighs in." I was fortunate that, in driving to my first ever live TV interview in Norwich, I heard a science correspondent discuss Frostie on BBC radio and give some useful sound bites on his significance for agriculture. Even with this coaching in how to be succinct and comprehensible, I was terrified as I spoke from a studio in Norwich to the presenter of the *Nine O'Clock News*, which had a vast audience in those days.

Thanks to my time with Chris Polge in the 1960s and 1970s, I had a background understanding of reproductive physiology, the basic techniques needed for this area of research, including surgery, and the management of experiments that were more likely to end in failure than in success. I began to think about life and death: when we attempted to revive the embryos from their frozen state, some were dead and unresponsive. But what was their status before they were thawed? And, of course, I began to appreciate the pressures of dealing with the media. This experience would prove useful when it came to Dolly.

ROSLIN, PHARMING, AND BEYOND

FROSTIE GAVE ME a reputation, one that presented new opportunities. In 1974 I joined the government-backed Animal Breeding Research Organization, ABRO, in Scotland. Initially I was based in the laboratories on the farm at Dryden, near Roslin, alongside the sheep sheds and pig barns, but in 1987 we were all moved as part of a major reorganization closer to the village of Roslin (*ross*, "a point," and *lynn*, "a waterfall"), where a ruined castle lies, a mining industry once waxed and waned, and the surrounding lowland hills were grazed—and still are—by flock upon flock of Scottish Blackface sheep. The village is also home of the fifteenth-century Rosslyn Chapel. Perched above the North Esk with its richly carved interior, it may contain, according to some believers, the Holy Grail, or the Ark of the Covenant, or part of the actual cross on which Christ was crucified.

What became the Roslin Institute in 1993 merged two research outfits, one concerned with poultry and the other sheep, cows, and other farm mammals. Today the Roslin is backed by the government's Biotechnology and Biological Sciences Research Council and inhabits a long, glassy building with a steep roof, mostly one-story—anything taller might be unwise, since the ground beneath is riddled with disused mining tunnels. The Roslin remains at the forefront of animal genetic research.

At first, I focused still further on the basic repertoire of reproduction—eggs, sperm, and embryos—with the aim of understanding the causes of infertility. I carved out a reputation as a research embryologist, and so when government funding went into reverse and the then ABRO concentrated on genetics, I was able to keep my job, despite being a physiologist. In the case of my institute, the aim was to cut our funds by two-thirds. The political dogma of the day was that scientific research must be utilitarian so that it could pay its own way, a philistine attitude that shows little understanding of the complex relationship between fundamental "blue skies" research (of the kind that usually seems pointless to a bean counter) and the development of innovative products. Under the prime minister at the time, Margaret Thatcher, research was being squeezed: on one side we were encouraged to do work that could have a payoff in the foreseeable future; on the other, any research that looked close to profitability should be cut adrift of government funding and be paid for by industry.

At that time molecular biology was becoming fashionable. The new director of the ABRO, Roger Land, brought my work on embryo death to a close in 1982 and told me to work on the genetic alteration of animals, then a tedious slog that usually ended in failure, but one that did draw on the skills of an embryologist. I had to end my research abruptly: either that, or resign. I was unhappy with the thought of merely recovering embryos from ewes and adding DNA and thought long and hard about leaving. There were then many redundancies, and the ABRO was under threat. Agricultural science in Britain felt almost doomed, and I discussed with Vivienne what I should do next.

The countryside intervened. Vivienne and I loved the Scottish borders. We played an active part in the local village, where I was a member of the farmer's club (of which I eventually became president). The surroundings were spectacular, and we walked our dogs in the hills. By now we had evolved into Lowland Scots: we decided to stay on, not least because we thought the children would benefit too.

I agreed to become a genetic engineer. I also told Roger that, once I had mastered genetic transformation, I would look for more efficient ways to do the job. Later that decade it was still a struggle (my colleagues at PPL between 1989 and 1996 took 2,877 ewes to create 56 viable lambs with new genes). I would come to realize that Roger wished to be involved in everyone's decisions, rather than give them the room to just get on with the job. As you may have gathered, I never saw eye to eye with him in the same way that I had with the relaxed and affable Chris Polge. Nonetheless, by the mid-1980s, I had become involved in the ABRO's effort to genetically alter a range of animals, such as mice, sheep, and pigs.

Genetic modification is, in one sense, as old as domestication. All varieties of domestic dogs, from Dalmatians to Chow Chows and Rottweilers, have been created through selective breeding—that is, breeding using only selected individuals. The same is true of cows and sheep. Now we wanted to find a way to alter the genetic makeup of animals with unprecedented precision. And we wanted to do more than merely improve the performance of farmyard animals. The idea was to give these animals an entirely novel role and convert them into living drug factories to make the human proteins factor VIII and factor IX to treat hemophilia and the enzyme AAT (alpha-1-antitrypsin), used to treat lung disorders such as cystic fibrosis. At that time this was a new field, a combination of pharmaceuticals and farming, aptly called pharming.

OLD-FASHIONED PHARMING

DURING THESE pioneering days, pharming was tiresome and repetitive. High science and lofty aspirations meant factory work and repetition, in reality. The idea was simple. Unfortunately, the practice was monotonous. Each gene in the body is, in effect, the recipe for a protein (it is, of course, more complex than this, since genes can be shuffled in various ways to make more than one protein). All one had to do was use a tiny hypodermic needle to introduce a gene into the meter or so of DNA packed into every cell. The way we did it—using a method first demonstrated with mice by Francis Ruddle and Jon Gordon of Yale University in 1980—was as crude and obvious as that. We squirted in naked DNA and hoped that it would be used by a developing embryo. By the time I started this work, the public had been alerted to the possibilities by one of the great pioneers in reproductive science, New Jersey–born Ralph Brinster, who had funded his education with a poultry business at the age of thirteen.

With his postdoctoral fellow, Bob Hammer, Brinster showed how one could make mice grow bigger—into "supermice"—by putting in a growth hormone gene. Then, with Ursula Storb, Brinster proceeded to turn implanted genes on in specific tissues, a handy trick if you want to make drugs in milk, for example. A recluse with a mustache who shuns most meetings and committees, and a Korean War veteran, Brinster has since done important work on sperm stem cells, on implanting them to restore fertility, and on their genetic modification. Hammer has remarked that Brinster's approach to complex scientific problems was always the same: "It's almost like he's waging war. He sees the problem, breaks it down, then goes after it. He's probably the most tenacious individual I've ever met."

After introducing genes into sheep embryos and introducing the resulting genetically modified (GM) embryos into sheep, around a thousand in all, I had a firsthand experience of how this new method of genetic modification was hopelessly and maddeningly inefficient. To do a thousand embryo transfers requires a military-scale operation with precision timing, day in and day out, to succeed. The methods of genetic modification we used in those days were not only inefficient but also limited: one could add a gene but not fix an existing faulty gene. Imagine the frustration of being a car mechanic who can only add new components to a faulty engine, never repair or replace them.

As I started this work, my own limitations became obvious. The dark art of embryo manipulation was beyond me. I had then, and still have now, a tremor that makes me hopeless at carrying out the tiny and controlled hand movements required to operate the microscope, at bending fine pipettes so that they can reach into the chamber in which the embryos are held, and at injecting genes—that is, if you could see the nucleus of the early embryo into which they have to be introduced.

Embryos are, in fact, opaque, but, thanks to painstaking observations by my colleague Paul Simons, we found that in seven out of every ten embryos we could find the nucleus by patiently rolling the embryo around and searching for a little clearing. Furthermore, there is not one nucleus but two. This feature of the earliest stage of life would confuse anyone who has read the school textbook account of conception. Look closely at a newly fertilized egg under the microscope, and you witness one of the common misconceptions about conception—within the early embryo, the egg's DNA remains separate from that imported by the sperm.

Within most cells the genetic instructions are often to be found in a package called the nucleus. But here, at the outset of a life, the DNA actually sits in two "pronuclei" (think of a farmhouse egg with two yolks). The existence of two nuclei, where female and male DNA sit in splendid isolation, raises a question about what we mean by an individual when we have an entity with a split genetic identity. In the pioneering days of *in vitro* fertilization, double nuclei led to heated debate in the scientific literature as to when the fertilization of test-tube babies really occurs. This debate has still not been resolved.

The good news for researchers in reproductive science is that this earliest stage of life contains lots of chemical machinery to put up with insults and errors. I suspect that the machinery includes some proteins that organize these pronuclei. We now rely on these proteins in cloning to reorganize the nucleus of a donor cell. This same machinery can also cope with foreign DNA, allowing genetic manipulation to take place. One may think that the foreign DNA should be put into both pronuclei. In fact, genetic engineers inject the DNA into one or the other. There is a limit to the insults that nature will tolerate, and to put new genetic instructions in both would be too disruptive and kill the embryo.

The results of this simple genetic manipulation are very much hit and miss. Of all the embryos injected with foreign DNA, only around 1 percent become live animals carrying a functional added gene. In fact, this low efficiency is more subtle than simple hit or miss, as I will explain in a later chapter. Nonetheless, we made steady progress. We produced our first transgenic lamb as the Brinster lab unveiled transgenic rabbits, sheep, and pigs in 1985. A company was started to produce in the milk of sheep human proteins needed to treat disease. From this work came our first breakthrough, Tracy, a Scottish Blackface sheep. Born in 1990, Tracy was probably the most famous sheep in the world until the birth of Dolly. As if to underline her commercial value, she was owned by our neighboring biotech company, PPL. Tracy's milk was rich in the human protein AAT, which was used to treat cystic fibrosis and emphysema. In every liter, she made thirty-five grams (to put this extraordinary feat in context, four of our other transgenic sheep managed to make only one gram or so). She was a living drug factory.

THE DRIVE FOR DOLLY

THE CREATION OF Tracy was, despite the effort involved, a relatively simple feat because it rested on adding a single gene. Most traits of interest to breeders depend on constellations of genes. Let's say you have a one-in-four chance of getting a working gene into a safe part of the genome (the complement of DNA carried by a person), a much higher success rate than currently possible. Imagine that you understand the biochemistry of a trait and that you want to "enhance" it by tinkering with ten genes. Then you would need about a million GM embryos to succeed. Traditional genetic modification was, to put it mildly, inefficient, when it came to altering complex characteristics.

And that is why I helped to create Dolly. Many people think that I was motivated primarily by cloning for its own sake, to take a superior animal and make identical copies of it. I did not seek simply to produce facsimiles of existing creatures, not least because this is impossible, as was highlighted above by the impact of serendipity, randomness, and chance on my upbringing. I am not concerned primarily with making multiples of elite livestock, though that was a motivation, and still less do I want to clone human beings. This has never been on my agenda. It was what other people, and the media, thought important. Cloning for me has always been a tool of science—finding out how cells work—and a means to improve the technology of genetic transformation of animals.

The original inspiration for the research to create Dolly dates back to work conducted early in the 1980s by Martin Evans and Matt Kaufman at Cambridge. They showed that it was possible to isolate lines of embryo stem cells from mice by means of so-called feeder cells, which provide the stem cells with support (in the form of a cocktail of as yet unknown cell growth factors and other trace chemicals). When an embryo is a few days old and ready to implant, it is hollow and contains a clutch of cells called the inner cell mass. Embryologists call it a blastocyst. The team found a way to grow and multiply cells from the inner cell mass of a blastocyst—the part that will turn into the fetus and most of the placenta.

These embryonic stem cells are rather strange. You can grow them indefinitely in the laboratory, and that provides lots of opportunities to genetically alter them. They also retain the unique ability to form all of the many different tissues that make up an adult animal. Mix these stem cells with the cells of an early eight-cell embryo or inject them into a blastocyst, and you find that the resulting offspring is a blend of the descendant cells of both the embryo and the newly introduced stem cells. Scientists call this blended creature a chimera (in Greek mythology, the chimera was a monster that had the head of a lion, the body of a goat, and the tail of a dragon). Despite this extraordinary potential, you can't grow a stem cell into a whole mouse, as you can with the very earliest embryo cells. This is partly because stem cells are too small to form an embryo, which is much bigger than other body cells. But you can grow them into particular cell types, such as nerve, skin, or bone cells for treatments, as we will see in a later chapter.

By blending stem cells with an embryo to make a chimera, one can turn the stem cells into the cells that make eggs and sperm: the mouse will then produce eggs and sperm that are the genetic offspring of the cultured stem cells. If you had genetically altered the embryo stem cells, the chimeric mice would also produce eggs and sperm with the alteration. This approach has been the basis of an extraordinary range of experiments in mice to study the role and function of newly identified genes. Because mice take relatively little time to become sexually mature—four to five weeks—the production of chimeras was fine when it came to creating GM rodents. But it would have taken us decades to achieve change in relatively long-lived creatures such as cattle or sheep.

The effort to make Dolly was born when it occurred to me that genetic modification would be so much easier and more efficient if we could first take stem cells from an early sheep embryo, multiply them to form millions of cells, then add DNA to those cells and select only the ones in which the process had succeeded for nuclear transfer to make animals. Whereas it is trivial to add genes to millions of cells in a test tube, then find the cells where genetic modification has worked properly, it would be impossible to achieve the same feat for classical genetic modification of millions of embryos. And it meant that, unlike the earlier scattergun approach, this method was also much less likely to produce sick GM animals (the industry was not going to benefit if the animals were hurt in any way by genetic modifications).

Only when you identify the embryonic stem cells in which your genetic transformation has been successful, do you go on to convert these GM cells into GM embryos. This final step is, of course, cloning. At the time I thought of doing this, however, cloning was only routine with bacteria and plant cells. I wanted to do the same with the cells of an animal. And to do this, I would have to extend an effort to clone a menagerie of cold-blooded creatures by a number of scientists that stretched back decades—an effort that had already stimulated an unprecedented amount of debate, controversy, and anxiety.

 TWO

A BRIEF HISTORY
OF
CLONING

DOLLY THE LAMB marked the culmination of more than a decade of intensive research at the Roslin Institute by myself and my colleagues. But the credit for this reproductive feat cannot be given to us alone. The achievement of her birth rested on a sturdy foundation of skills, understanding, and invention that had gradually been laid by scientists across the world since the nineteenth century. And it relied on a zoo of species. The early clones that blazed the trail to Dolly were not just sheep but also rabbits, mice, and cattle. Perhaps the most important pioneers of all were cold-blooded clones of sea urchins, salamanders, frogs, and toads. Despite this long history, the path to Dolly was far from straightforward. There were distractions caused by mistakes, muddles, and misunderstandings. There were rivalries and controversies, even accusations of deceit. All the while, many speculated on the implications of advances in reproductive science. These ideas were, in turn, embellished in novels, movies, and other imaginative creations. Fictional fascination with cloning has rarely focused on scientific fact but usually on issues of identity and how the sanctity of life will be challenged when "ditto machines" of one kind or another create "cookie-cutter humans." This obsession has led to endless confusion about what is possible and what is not.

While some people fear (wrongly) that a clone would be a duplicate not just of a person's DNA but of his or her consciousness too, others fear (wrongly) that clones would be soulless replicas. Popular culture has become obsessed with the idea that, in a world where individuals are easily duplicated by cloning, life will be cheap and a person will be reduced to a replaceable unit, a mere machine part of society.

Yet, in one respect, science fiction is already science fact. Cloned humans have been around since the dawn of humanity. A few days after an egg has been fertilized by a sperm and reaches the stage of the blastocyst, it has to "hatch" from its rubbery "shell," the zona, so that it can implant in the wall of the uterus. During this process one individual can become two: the embryo can split. Sometimes the process of twinning does not occur properly, leading to conjoined twins. One striking, and ancient, depiction is a marble statue of two women linked at the hip, excavated from a Neolithic shrine in Anatolia. Down the ages, great minds have pondered what conjoined twins tell us about development. The Greek polymath Aristotle (384–322 BC) mused on whether they arose from two embryos that had combined or from one that had split: fusion or fission?

Intriguingly, we now know that both can occur in the womb. Rarely two eggs ovulate, and both can become fertilized. And sometimes those embryos can fuse to create a chimera. Alternatively, an embryo can undergo fission to create two. There are variants on the theme of identical twins, depending on the degree of fission and how much plumbing they share in the womb. Some twins still draw on the same placenta. Others will develop with separate placentas in different parts of the womb. Of those, some will have their own membranes and others share them.

Embryo cleavage happens in the rest of the animal kingdom too. The nine-banded armadillo produces four genetically identical progeny from each fertilized egg. This "natural" process of fission was also exploited by scientists' very first attempts to create a clone, before the other type of cloning—by the process of nuclear transfer—was developed to do the job, the first creation of a normal fertile adult animal from a body cell of another adult being the most dramatic example of all.

PIONEERS OF CLONING

EARLY CLONING used critters that, in a way, have been much more important and influential in this field than sheep: sea urchins, salamanders, frogs, and other creatures that have eggs so big that they are easy to study. They offer another advantage: whereas the Dolly effort relied on a labor-intensive combination of hormones and surgery to harvest a few eggs from a ewe for cloning, frogs make fabulous quantities—thousands and thousands—and deliver them in the convenient form of spawn. In short, frogs and their amphibian friends are gifts for the curious cloner.

One of the first to try embryo splitting was the German Hans Driesch (1867–1941), the son of a gold merchant who became a philosopher, biologist, and master experimentalist. He took a glassy two-cell sea urchin embryo and shook it apart. Both cells developed into complete individuals. Then he dismantled a four-cell embryo, and quads were the result. This work marked probably the first successful human-assisted animal cloning from a single embryo. In some ways he set the pattern for future work. Driesch, like me, was driven by an urge to explore basic science, not clone for its own sake. Like me, he was in awe of development in which a fertilized egg divides again and again, differentiates, and becomes more complex, as cell layers slide over each other and fold in upon themselves in highly choreographed auto-origami to create an urchin, a frog, or a man. Unlike me, however, Driesch was convinced that the ultimate explanation for development lay outside the boundaries of science: he believed that a "vital force" lay behind the miraculous transformation from fertilized egg to adult.

Driesch was the last great spokesman for this idea of vitalism. Though it still seems miraculous, we now understand much more about the origins of a life. Like other developmental biologists, I see no evidence of an invisible hand at work. The laws of physics and chemistry are sufficient to explain the behavior of the molecules in a living cell. There is no room or need for a vital force.

The pioneers of cloning were trying to understand the basic details of the biology of development. These first forays into the field focused on whether humans were fully formed at the moment of fertilization, and just grew larger, or whether they developed from a single cell to the complex arrangement of billions of cells in a person. By the time Driesch entered the story, there were two conflicting theories on how cells gradually differentiated from the blank slate of an embryonic cell to have a dedicated purpose, whether as a brain cell or a liver cell: the cells either lost all genes, except those relevant to their specific task, or they kept the entire genetic recipe—genome—and selectively turned genes on or off.

The experiments of Driesch were regarded as refutations of a then prevalent idea of another great German, August Weismann (1834–1914), of the University of Freiburg. In the 1890s, Weismann went on to challenge Jean-Baptiste Lamarck's theory that traits acquired by parents during their lives could be passed on to their offspring (a theory overtaken by those of Charles Darwin and Gregor Mendel) by cutting off the tails of twenty-two generations of mice (1,592 in all), demonstrating that the injury was not passed on, and he also became an influential figure in the history of eugenics.

Weismann believed that the genetic information of a cell would peter out as the cell went through differentiation. If he was right, separating and growing each cell from a two-cell embryo could create only half a creature. But the embryo-splitting experiments that Driesch conducted at the end of the nineteenth century show how, a few divisions after fertilization, embryonic cells retain the ability to turn into any type, from heart cells to egg or sperm cell, or even a whole individual (biologists say that they remain totipotent).

In the wake of subsequent work, we now know that single cells taken from sheep embryos at the stage when they have two cells are each equally able to become lambs. The same goes for other livestock. One might expect that the calves or lambs that result from fission would be smaller than usual, but recent work has shown they are actually normal in size. Part embryos enjoy a growth spurt to get back on the normal developmental track. Conversely, when embryos are lumped together, the rate of growth declines accordingly.

This form of cloning, which has a long history, suffers from the law of diminishing returns: there is a limit to the clones that can be divided out of a single embryo, and this limit varies from species to species. The leading British embryologist Anne McLaren, while working in London with colleagues, found it a struggle to obtain pups from mouse embryos at the two-cell stage that had been divided in half at the two-cell stage.

This method has been used to duplicate our close relatives. A rhesus monkey was the first to be cloned this way, in research by Gerald Schatten while at the Oregon Regional Primate Research Center. He took rhesus monkey embryos at the eight-cell stage and separated them into four two-cell groups, creating a group of identical quadruplets. Two embryos were put into each of two surrogates. Both became pregnant, each with only one of the two embryos. One miscarried and the other resulted in Tetra.

The same method could be used on people too. A much misrepresented experiment using early human embryos in 1993 by Jerry Hall and Robert Stillman of George Washington University suggested that fission could be used to boost the numbers of embryos, and thus the chance of success, of *in vitro* fertilization. Although this type of cloning is "natural" and their aim was reasonable, the research caused a furor. This application is unlikely to catch on because of public fears stirred by the "c" word.

NUCLEAR CLONES

THE FIRST GLIMPSE of the nuclear transfer process that we used to create Dolly came decades after the first attempts at twinning in the laboratory. One element of nuclear transfer arose from the German-American Jacques Loeb (1859–1924). While working at the University of Chicago, Loeb showed how to simulate fertilization in sea urchin eggs, a process we now call activation. Another element came from the pioneering and innovative German embryologist Hans Spemann (1869–1941), son of a publisher and, among other things, director of the Kaiser Wilhelm Institute of Biology in Berlin. In 1919 he was appointed professor of zoology at the University of Freiburg im Breisgau, a post that he held until he retired in 1935. During his influential career, Spemann helped establish what he termed *Entwicklungsmechanik* (what we would call developmental biology). Awarded the Nobel Prize in 1935, he was the only embryologist to get the honor for the next half century.

With Hilda Mangold (who died at the age of twenty-six, after her dress caught fire while she was refueling a kerosene stove), Spemann revealed remarkable details of development by transplanting tissue between newt embryos to create conjoined-twin newts. He also conducted the first nuclear transfer experiments on the newt, once again a favorite subject for this kind of work because of its large and thus easy to manipulate eggs. For aficionados Spemann is the true father of cloning.

In the winter of 1896–97, while convalescing from tuberculosis in a sanatorium, Spemann read Weismann's theory of heredity and development and was inspired to investigate. In 1914 he published an account of a fascinating experiment. His enterprise culminated in 1938, with his masterwork, *Embryonic Development and Induction*, which described his experiments with a rudimentary form of nuclear transfer.

Key to his technically astonishing technique was an unusually simple piece of equipment: a delicate hair from a blond infant less than nine months old (after his death, a lock from Margrette, his baby daughter, was found in an envelope, tucked in his 1899 file holder). Working on the slippery eggs of salamanders, hundreds in all, while hunched over a binocular microscope, he used the hair as a noose to constrict a newly fertilized salamander egg and was able to divide the cytoplasm of the egg—the part inside the cell membrane and outside the nuclear membrane—so that he created two sections of the embryo. In one section of the resulting dumbbell-shaped embryo, there was a nucleus—containing DNA—in the other, there was only cytoplasm, or other cellular material. After the nucleated side divided four times, creating a sixteen-cell embryo, Spemann loosened the hair and allowed one of the sixteen nuclei to slip back into the separated cytoplasm.

Spemann had created a new cell made up of some of the original egg fluid and a new, more mature nucleus. Cell division now began on this side as well, and by tightening the hair loop again, Spemann broke apart the two embryos. It was an extremely elegant experiment. Spemann had moved the original nucleus out of a portion of an egg and then put a more developed nucleus back in its place. He had shown that the nucleus retained the ability to turn into any type (was totipotent) after undergoing four divisions. He had invented a new way of cloning. What a neat trick! All the more so, given that it was accomplished with two tweezers and a baby's hair. This experiment had an astonishing payoff: a twin set of salamanders, one moments younger than the other. Spemann called it twinning, but we know it today by a different, more loaded word: "cloning." Spemann had documented the first *in vitro* animal clone produced by nuclear manipulation.

He wanted to go much further. In his 1938 book Spemann proposed a "fantastical experiment," a *gedanken* (thought) experiment that would lay the basis for Dolly. It would be salutary, he said, to take the nucleus of a differentiated cell—even from an adult cell—and place it in the cytoplasm of an egg whose own nucleus had been removed (enucleated, as we say today). But he did not do this experiment, for practical reasons: although he could work out how to isolate a nucleus, he did not know how to introduce it into an egg that had been drained of its own DNA—how to carry out what we now think of as nuclear transfer. For the science of his time, this proposal really was fantastical. It was as provocative as asking whether we could reconstruct a person after ripping the brain from one body and thrusting it into the emptied skull of another. Indeed, Spemann himself was so in awe of development that he suspected that it depended on more than mere physics and chemistry.

Even though we know and can do so much more today, Spemann was truly remarkable and exceptional. He was working by hand on small eggs, whereas we use instruments that can give precisely controlled small movements. Not only was he an experimental genius, he was thinking differently. After all, he had the basic idea for the Dolly experiment. In the next decade, Spemann's challenge caught the eye of Robert Briggs, who had worked in a shoe factory and also earned money as a banjo player in a dance band before a high school teacher spurred his interest in biology. He later recalled, "At the time it never occurred to me that I would become a biologist. I didn't even know that one could earn one's living this way."

By 1949 Briggs was a researcher at Lankenau Hospital Research Institute in Philadelphia (later the Institute for Cancer Research and now the Fox Chase Cancer Center). He was studying the role of the cell nucleus and its cargo of chromosomes in development: What governs how the genes on a chromosome are used by an embryo cell to lay down its identity? A colleague, Jack Schultz, suggested he try an echo of Spemann's fantastical experiment: nuclear transfer could probe whether genes are turned off during development or whether they are lost forever.

When Briggs applied to the National Cancer Institute to carry out nuclear transplantation, his proposal was deemed a "hare-brained scheme" with little chance of success. He reapplied and this time, after a site visit by an institute representative, the application was approved with enough money to hire a dextrous assistant—Thomas King of New York University, an expert on micromanipulation who was at that time working all hours to support his family.

Together, they did their studies on the common American leopard frog and its big, one-millimeter eggs. Using microscopic pipettes, needles, and watchmaker's tweezers, they emptied the egg of DNA and transferred into it the nucleus of a frog cell taken from an early embryo. These delicate manipulations paid off in November 1951 when they successfully reconstructed frog embryos that formed tadpoles. Nuclear transfer "may have other uses," they speculated in the *Proceedings of the National Academy of Sciences* in March of 1952. This failed to anticipate all the hype, activity, and headlines that followed the arrival of Dolly. Nevertheless, this anodyne statement was enough to electrify the biologists of the day.

Briggs and King examined the question raised by Spemann—could an adult cell retain the ability to direct embryonic development? They found that about half of the nuclei from blastocysts generated normal tadpoles, showing that their method worked in principle. But, even though it was no more technically demanding, their attempt to use nuclei from the more differentiated cells in the next stage of embryonic development from the blastocyst, the gastrula, did not work. (One reason is that a fertilized egg divides rapidly compared with the cell cycle of the transplanted nucleus, which winds down as it differentiates.) They concluded that developmental potential diminishes as cells differentiate, and that it was impossible to produce a clone from the nucleus of an adult cell. The implications for any proposal to clone from an adult cell are clear: Dolly was, by the understanding of the 1950s, a fiction, a mirage, an impossibility.

Evidence that all of the cells of an adult contained the entire recipe to make an individual—the genome—was obtained in the late 1950s by John Gurdon, a graduate student at Oxford University working under the supervision of Michael Fischberg, a Swiss developmental biologist. That Gurdon even began a career in biology at all is remarkable. While at Eton, he had been told that science was not his strong point. In fact, his schoolmaster delivered a withering verdict on Gurdon's hope of becoming a scientist: "This is quite ridiculous, if he can't learn simple biological facts he would have no chance of doing the work of a specialist, and it would be a sheer waste of time both on his part and of those who have to teach him." Despite this devastating judgment, the time waster found himself as a mere graduate student challenging the wisdom of Briggs and King, then established figures in the field.

Instead of leopard frogs, Gurdon was trying to clone the African clawed frog *Xenopus laevis* (*Xenopus* is Latin for "peculiar foot," an apt description of the enormous webbed, five-toed, three-clawed rear feet; *laevis* means "smooth"). *Xenopus* eggs are about one-tenth the volume of those of leopard frogs and have elastic membranes that make them even harder to work with. But they were easy to persuade to make eggs—they had to be injected late the night before the eggs were needed (this research was not good for Gurdon's social life). And they were plentiful in the 1950s because female *Xenopus* frogs were used in a hospital pregnancy test in which a female *Xenopus* would make eggs if injected with the urine from a pregnant woman. The scientific question Gurdon asked of these frogs when conducting thousands of transplants of cell nuclei was whether the nuclei from these differentiated cells were fully potent—totipotent—and thus able to reconstruct an entire organism.

Gurdon, now Sir John in recognition of his outstanding contribution to science, showed that this was indeed the case. He carried out his experiments with tadpole intestinal cells, including some that were differentiated. In 1966 he reported that when he popped the DNA of a cell nucleus from a juvenile animal into enucleated eggs, he could make one frog every so often. "In all, 120 transplanted intestinal nuclei yielded seven adult frogs, five of which were fertile," Gurdon later wrote. "That experiment was perhaps the single most important one we performed, because it proved that a cell can undergo specialization and yet remain totipotent, retaining all the genetic material needed to make a complete, sexually mature individual. This key experiment justified the view that the cloning of differentiated, and perhaps even adult, cells was at least theoretically possible."

But questions were raised about these experiments. They went far beyond what the work of Briggs and King suggested should be possible, and the *Xenopus* results were thus greeted with skepticism. Some scientists were concerned that a small but significant percentage of intestinal cells used by Gurdon for cloning might have contained relatively undifferentiated stem cells, because they originated during development in the endoderm, the part of the early embryo that turns into the gut. Others, for reasons that Gurdon himself still does not understand, suspected that germ cells could have been present in the intestine and that these undifferentiated cells enabled his experiments to succeed. In short, people doubted that it was possible to clone with cells that were more aged than those used by Briggs and King.

These criticisms spurred the Oxford team on. In 1975 Gurdon, Ronald Laskey, and Raymond Reeves reported in the *Journal of Embryology and Experimental Morphology* the transplantation of the nuclei of skin cells from foot webs, providing convincing evidence that Gurdon really was using fully differentiated donor cells. They used a marker, an antibody, to convince the critics that the donor cells were indeed skin cells. Around this time he also took a spectacular photograph of thirty little albino frogs, cloned by nuclear transfer from the cells of an albino tadpole into the eggs of normally pigmented frogs. The image appeared in the newspapers. Some readers must have been aware of the echoes of Aldous Huxley's *Brave New World*, in which "Bokanovsky's Process" could turn each fertilized human egg into ninety-six perfectly formed embryos. ("'But alas,' the Director shook his head, '—we can't bokanovskify indefinitely.'")

At this point you may be wondering how these experiments fell short of the success that the Roslin team later claimed with Dolly. Perhaps Gurdon's studies lacked the glamour of a mammal experiment. Perhaps, if Gurdon had called his first albino clones Whitey, Footy, or even Webster, he could have captured enough publicity to rewrite cloning history. But, more important, Gurdon's experiments were never able to generate an adult animal from the nuclei of cells derived from an adult animal. He had produced fertile adult frogs from the nuclei of tadpole intestinal cells, but those cells came from a *juvenile* rather than an adult animal. Although he had grown heartbeat-stage tadpoles from the nuclei of adult skin cells, none of his skin cell embryos survived to become adult frogs. As a consequence of all of his studies, he concluded that transferred nuclei from cells that are more specialized are less likely to support development beyond the early stages than are less specialized cells. Although he did not clone an adult frog from an adult frog cell, his important work did underline that the genome remains intact during the course of cell differentiation, a key principle, and nicely complemented what was to come with Dolly.

Today Gurdon comes across as an archetypal old-school Englishman. Well-spoken, with floppy brown hair and a love of fast cars, he is a man of independent means and blessed with an indifference to mortgages and other tiresome features of everyday financial survival. He has an acute mind and, as important, the ability to ask the right questions. Above all else, he remains passionate about his science. Despite his age and his eminence, Gurdon can still be found working at the laboratory bench in the research institute in Cambridge that now bears his name.

NUCLEAR RABBITS AND MICE

THE STORY OF mammalian cloning can be traced back to one of Gurdon's pupils, Chris Graham, a chain-smoking intellectual with a slow turn of phrase and a dry sense of humor. Inspired by Gurdon's success with frogs, Graham had moved in the mid-1960s to Oxford University, where he worked in the department of Henry Harris, a Renaissance man. Harris was born and educated "under a hot Australian sun," first read modern languages and then turned to medicine, though he remained interested in history and writing, whether fiction or a vivid account of the science of the cell.

Harris had come up with a key insight into cancer after he had fused cancer cells with normal cells and found that the resulting hybrids grew normally, evidence of what are now called tumor suppressor genes. Using Harris's method of fusing cells with the noninfectious Sendai virus, Graham wanted to develop a gentle nuclear transfer technique for mice.

The virus is able to link together the outer membranes that protect cells for a short while, like the skins of merging soap bubbles. With two nestling side by side, the membranes flow together and fuse to enclose the two nuclei together to create hybrids. With John Watkins, Harris had published a 1965 paper in *Nature* describing how they had even fused human and mouse cells to form "heterokaryons" with genetic material from both species. This technique opened up limitless possibilities for the study of the relationship between the nucleus of a cell and the cytoplasm, cell signaling, and the function of genes. The media were, of course, more intrigued by the possibilities of man-mouse hybrids, the stuff of nightmares.

Since then scientists have indeed used the virus to fuse unlikely couples such as insects and mammals for basic research. The virus also seemed to offer Graham a way of easing a nucleus of a cell into an egg without the trauma of an injection needle. The latter threatened to disrupt a mouse egg, which is only one-seventeenth the diameter of a frog egg and just one five-thousandth of the volume.

Working in Oxford's Dunn School of Pathology, Graham used the virus to pass a variety of cell nuclei into the one-cell stage (zygote) and two-cell stage embryos of mice. However, he found no evidence that the implanted nucleus participated in the development of any embryo. In fact, even if the new nucleus had taken part in development, there would not have been any live births: the result of these experiments was always genetically abnormal, with double the usual number of chromosomes. Nonetheless, Graham felt that he was on the path to mammalian cloning.

Today, after a long career that included important work on the genetic alteration of animals, Graham recalls how he wrote "an outrageous paper" for a 1969 symposium in Philadelphia in which he indicated that successful nuclear transfer in mammals was just around the corner. He explained that he could enucleate eggs without damage (confirming earlier work by the IVF pioneer Robert Edwards) by using an alkaloid from the autumn crocus (colchicine) that confuses cell division. He described how the egg could then be parthenogenetically activated (tricked into dividing without the help of a sperm) and, referring to his earlier efforts to fuse cells with Sendai, suggested that successful mammalian nuclear transfer was only a matter of blending these methods in one experiment. This happy event escaped him, partly because he moved the eggs and cells together with a handheld pipette, without the help of a specialized instrument called a micromanipulator, and partly because scientists were still developing reliable methods for keeping a single-cell embryo alive after it has suffered this kind of physical abuse.

Derek Bromhall, another of Oxford's graduate students, would build on the work of Gurdon, Graham, and others. Fascinated by the possibilities of transferring Gurdon's frog work to mammals, Bromhall wanted to give up his career as director of fisheries research in Hong Kong to investigate the possibility of nuclear transplantation in mammals. The chief scientific adviser to the UK government, Solly Zuckerman, had visited his lab in Hong Kong and backed Bromhall's idea, even though many were doubtful that the feat was possible, given that mammalian eggs are much smaller than those of a frog—in fact, smaller than the period at the end of this sentence.

Bromhall moved to Oxford in 1967 to begin his work with rabbits. Although their eggs are larger than mouse eggs, rabbits are more prone to infection, and surgery must be done in sterile conditions. Over the next seven years, Bromhall developed microsurgical equipment and culture techniques required for nuclear transfer, attempting to synchronize the cell cycles of egg and donor nucleus (by incubating the donor in an atmosphere of nitrous oxide under pressure to arrest the cell cycle) and enucleating eggs by means of gamma rays or UV irradiation to destroy the DNA (the microsurgery of the day destroyed the eggs). He ran parallel experiments with the Sendai virus and the microinjection of a variety of cells, but concentrated on transferring nuclei from early rabbit embryos (at what embryologists call the morula stage, the ball of cells that forms before the blastocyst) to unfertilized eggs by microinjection. At that time there were no antibody markers, as there were in Gurdon's frogs, and it was essential to prove that the cells of an embryo created by nuclear transfer were from the transplanted nucleus; otherwise it could be argued that the embryo had developed parthenogenetically, which is possible for a while.

In an attempt to follow the fate of the transplanted nucleus, Bromhall labeled the donors with radioactive thymidine (thymidine is involved in the synthesis of DNA and in the preservation and transfer of genetic information). Unfortunately, this technique necessitated killing any embryos created by nuclear transfer by slicing them into thin sections, which were then coated with a photographic emulsion. Exposed for several days and then developed, the nuclei in the embryo's cells derived from the transplanted nucleus could be identified by clusters of black spots caused by the radioactivity in the thymidine.

The next step would have been to transfer the cloned embryos to the uterus of a recipient rabbit. It was at this stage that the last of Bromhall's research grants, from the Cancer Research Campaign, ran out, and Bromhall changed careers again, becoming a producer of documentary films. I doubt that he would have been successful. However, at that time very few of the enabling techniques that we took for granted in the creation of Dolly were well established, so when Bromhall reported this work in 1975 it was remarkable enough to grace the pages of the journal *Nature* on Christmas Day. This was a milestone, the first published attempt at cloning a mammal by nuclear transplantation.

During Bromhall's thesis work, Gurdon had left Oxford. But the reputation of the lab still attracted visits by key figures in this developing field such as Andrzej Tarkowski of Warsaw University. Tarkowski had used electricity to persuade mouse eggs to divide, had studied the detailed effects of the egg on an alien nucleus, and later went on to use electricity to fuse mouse embryos. Bromhall's work was "not very convincing," according to Tarkowski, who today is a professor in the classic central European style, a serious man with a passion for biology and a talent for photography.

Tarkowski thought that one of his colleagues, Jacek Modliński, had in 1978 provided stronger evidence that nuclear transfer worked in mammals when he injected embryonic nuclei into nonenucleated mouse zygotes and witnessed the development of chromosomally abnormal embryos to the blastocyst stage. "He presented undisputable proof that the injected nucleus was incorporated into the genome (genetic makeup) of the operated embryo," recalled Tarkowski.

While he was trying to clone his rabbits, Bromhall was also visited by Davor Solter, a scientist born in Yugoslavia (now Croatia). Then in his midthirties and based at the Wistar Institute, in Philadelphia, Solter wanted to find out more about the possibilities of mammalian cloning. Although Bromhall's work was pioneering, Solter was struck by what a struggle it was to get his methods to work. "I watched Derek Bromhall work for a week without a single successful transfer." There was still much left to be done. But not in Oxford. This early powerhouse of nuclear transfer would become more preoccupied with molecular biology. A decade later, as mammalian cloning moved to other laboratories, Solter became a significant player in the field.

IN HIS IMAGE

THESE EARLY cloning efforts were used as the inspiration for a book by David Rorvik, *In His Image: The Cloning of a Man*. Originally billed as the true story of the cloning of a mysterious millionaire, the novel was published in 1978, the year that the first test-tube baby was born. With a readership that had been inspired by the possibilities raised by that momentous birth, his book caused a sensation, generating many column inches of comment in newspapers, a congressional hearing, and endless angst.

This is how Rorvik's story goes: a self-educated, aging millionaire businessman named Max yearned for a son. He asked Rorvik to use his connections to recruit scientific talent for a secret project to clone an heir from his own DNA. In return he would allow Rorvik to write about the reproductive wonder in the making. According to Rorvik, a team was assembled, headed by a brilliant but overweight scientist code-named Darwin. They were flown to and set up in a laboratory on a secret island somewhere in the Pacific. After years of experimentation, they succeeded in achieving a pregnancy in a surrogate mother, a local teenager code-named Sparrow. Nine months later, she gave birth to the first human clone. Rorvik wrote, "It was not, I thought, exactly the nuclear family. But it was a thrilling sight, this old man, this young girl, this strange baby." He insisted that his work was nonfiction, but the book was printed with a disclaimer, informing readers that it was for them to decide. Rorvik refused to divulge details "to protect the child from harmful publicity." It is touching to see a journalist express such concerns.

Scientists cried foul, pointing out, among other things, that no one had ever grown a frog to adulthood, let alone a mammal, from the specialized cells of an adult. But the book became a best seller. And politicians took notice. Briggs had testified at the 1978 hearing in the House of Representatives on Rorvik's sensational book that "cloning in man or any other animal is not just a technical problem to be solved soon but may, in fact, never occur." They had their own experiments to draw on as evidence of this fact. But perhaps they had done something wrong. As Briggs and King later wrote, "at the time we could not eliminate the possibility that the developmental deficiencies might be the result of nuclear damage during the operation."

To give his work a veneer of respectability, Rorvik had cited Derek Bromhall's efforts to clone rabbits. This backfired. Bromhall was infuriated. He remembered that Rorvik had written to him some time before, explaining how he was writing a review of nuclear transplantation, and, "assuming that Rorvik was an honest researcher," he had sent him the abstract of his doctoral thesis. Bromhall would sue the publishers, J. B. Lippincott, for dragging his name into the furor.

He had never consented to the inclusion of his name or research techniques in the book and did not want to be associated with the idea of advocating human cloning. In 1979 a judge ruled that *In His Image* was a hoax. Three years later the publisher admitted that the book was fiction.

Bromhall himself was no stranger to fictional cloning. He was the scientific adviser for *The Boys from Brazil*, which linked cloning with totalitarianism. Duplicates of Adolf Hitler starred in this 1978 movie version of the Ira Levin novel of the same name. It is an unfamiliar take on the familiar script about Nazis trying to rule the world. Hitler's mass executioner, the geneticist Josef Mengele, having escaped to Brazil, clones ninety-four boys from Hitler's blood and places them with adoptive parents. The boys are, of course, little rotters. But even Nazis know that having the "right" genes isn't enough to grow an authentic Adolf. After all, the real thing might have painted landscapes for the rest of his life had he not been rejected by the Vienna Academy of Fine Arts.

The clones are placed in adoptive homes in North America and Europe and raised in a way that roughly resembled the führer's own upbringing. As each proto-Hitler nears the age of fourteen, his father dies, as did Hitler's own father when he was that age, thus re-creating the microculture for the clones that would allow each mini-Hitler to become evil incarnate. In this way the film tapped the deep unease about perverting science for political and ideological ends.

Intriguingly, Bromhall made a film insert illustrating the techniques of nuclear transplantation to explain how the cloned Hitlers could have been created. He would himself go on to produce wildlife and medical documentaries, for one of which, following the growth of a baby from conception to birth, he was nominated for an Oscar. With the cooperation of the staff at the Roslin, Bromhall also documented the story of Dolly until the birth of her first lamb, Bonnie.

BRAVE NEW WORLD

As NUCLEAR TRANSFER was conducted on amphibians and then applied to mammals, eminent scientists were also inspired to think through the implications. Take, for example, the great geneticist J. B. S. Haldane, who helped create the cultural backdrop to cloning. His 1924 book *Daedalus* explores the supposed advantages of "ectogenesis," the growing of human embryos in artificial wombs. In this way Haldane influenced Aldous Huxley's 1932 novel, *Brave New World*, in which lower castes of a dystopian society are cloned to serve totalitarian ends.

Three decades later, in 1963, Haldane gave a remarkable speech entitled "Biological Possibilities for the Human Species in the Next Ten Thousand Years," which introduced the term "clone" to a wider audience. His views were born of a naïve confidence and optimism about science that was prevalent at that time. Oblivious of some of the darker implications, they were driven by excitement at the thought of cloning a Mozart, rather than cloning an embryo of unknown potential (let alone one of the hoi polloi). Haldane told his audience,

> Perhaps the first step will be the production of a human clone from a single fertilized egg, as in *Brave New World*. But this would be of little social value. The production of a clone from cells of persons of attested ability would be a very different matter, and might raise the possibilities of human achievement dramatically. For exceptional people commonly have unhappy childhoods, as their parents, teachers, and contemporaries try to force them to conform to ordinary standards. Many are permanently deformed by the traumatic experiences of their childhoods. Probably a great mathematician, poet, or painter could most usefully spend his life from 55 years on in educating his or her own clonal offspring so that they avoided at least some of the frustrations of their original.
>
> On the general principle that men will make all possible

mistakes before choosing the right path, we shall no doubt clone the wrong people. However everyone selected for this purpose will presumably exceed the median considerably in some respect, if only as a humbug. And the greatest humbugs, like Hitler, would hardly relish the thought of producing a dozen possible successors with their own abilities, and youth to boot. Possibly a movie star at the age of forty might have similar feelings.

Assuming that cloning is possible, I expect that most clones would be made from people aged at least fifty, except for athletes or dancers, who would be cloned younger. They would be made from people who were held to have excelled in a socially acceptable accomplishment. Sometimes this would be found to be due to accident. The clonal progeny of Arthur Rimbaud, if given favourable conditions, might have shown no propensity for poetry, and become second-rate empire builders. Presumably such a clone would not be further grown. Other clones would be the asexual progeny of people with very rare capacities, whose value was problematic, for example permanent dark adaptation, lack of the pain sense, and special capacities for visceral perception and control. Centenarians, if reasonably healthy, would generally be cloned, if this is possible; not that longevity is necessarily desirable, but that data on its desirability are needed. Centenarians who could continue to learn systematically up to the age of thirty would almost certainly be useful, and probably happy members of society.

Haldane, who died the next year, was not the only scientist to warm to the possibilities of cloning during this age of wide-eyed optimism. The Nobel laureate Joshua Lederberg also discussed how to use cloning to "improve" mankind in an article in the *American Naturalist* in 1966. Others were not so sure. Four years later came Alvin Toffler's influential warning about the disorienting pace of scientific and technological change, *Future Shock*, which, like Haldane, anticipated the story line in *Boys from Brazil*. Toffler mused on how there was a certain charm to Albert Einstein's bequeathing copies of himself to humanity, "but what of Adolf Hitler?"

Haldane's own sister Naomi Mitchison further underlined the naïveté of the initial discussions of cloning in her 1975 novel *Solution Three*. Her title was meant to echo the Nazis' Final Solution, the drive for racial purity that led to the horrors of the death camps and human experimentation. This sci-fi novel had a more compelling provenance than most: it was dedicated to the DNA pioneer Jim Watson, who first suggested the idea. Another key influence was the embryologist Anne McLaren (who, incidentally, had appeared as "The Child" in Alexander Korda's 1936 sci-fi film *Things to Come*). And Mitchison's childhood friends included Julian Huxley, who would become an eminent zoologist, along with his younger brother, Aldous.

Mitchison populates her novel with lesbian, gay, and straight characters who live in an emancipatory society based on mandatory homosexuality, banned heterosexuality, and a new model for reproduction using "Clone Mums" and systematic "conditioning" of their cloned children. *Solution Three* is driven by the need to cope with rising world population and falling food production, as efforts to create a narrow range of superplants have left major crops vulnerable to disease.

Mitchison explores the feminist temptation to use cloning on human guinea pigs to wipe out sexism, racism, war, and other trappings of an oppressive, diverse, and patriarchal society to produce instead a uniform, predictable, and peaceable society, a forest of human genetic excellence. Her novel reminds us about the importance of nurture, not just nature, the value of diversity, and the unexpected costs of even the most well-intentioned technological fix. She ends on an optimistic note that flexibility, not dogma, can win out.

Today her older brother's speech fascinates and chills in equal measure and reminds me of one of the first things I learned at university—there is no clear relationship between intelligence and a sense of social responsibility. Because of this I find it understandable that the media still use *Brave New World* as shorthand for the direction in which the scientific control of reproduction could take us without thinking through the implications and subjecting them to intense public scrutiny.

A FALSE START

ALTHOUGH VERY RELEVANT to the cloning of mammals and people, the early work that so inspired Haldane and Lederberg must be set into its proper context. Frogs and salamanders are amphibians. They are easy to work with because their eggs are huge compared with our own—human eggs are minute by comparison, as ping-pong balls are to pumpkins. In addition to being measurable with a school ruler, the eggs of amphibians are easy to get at. As easy as collecting frog spawn in a jam jar. But mammal eggs never see the light of day. They are hidden deep in the female body in various types of plumbing such as the uterus and oviducts. At the very least, many new techniques would have to be developed if this pioneering work with amphibians were to be applied to mammals. Despite the pioneering work in Oxford, Warsaw, and elsewhere, an understandable suspicion was prevalent among scientists that there were profound obstacles. Many scientists expected, quite understandably, that the hurdles would be huge, and some feared that it might never be possible to clone mammals at all.

That view was challenged by some truly fantastical research on mice. Mammalian cloning seemed to be launched in earnest with the experiments of the American Peter Hoppe and Karl Illmensee of the University of Geneva, a trim and dashing German embryologist thought to be something of a wunderkind. In the late 1970s, at a scientific gathering known as a Gordon Research Conference, they astounded their peers by claiming to have created mice with all-male or all-female parents. In a 1977 paper in the *Proceedings of the National Academy of Sciences*, they described how they had done this by removing a female or a male pronucleus from an unfertilized egg. They made up for the loss of genetic material by doubling (diploidizing) the genetic material in the remaining pronucleus with the help of a fungal metabolite called cytochalasin, which interferes with cell division by inhibiting the replication of the proteins that form the cellular skeleton. We now know, because of the influence of a phenomenon called imprinting (described later), that this is a dubious proposition and that the resulting embryos do not develop beyond midpregnancy.

Then, in January 1981, Illmensee and Hoppe announced in the journal *Cell* that they had cloned three mice. These mice were all the more remarkable because a succession of failed attempts was beginning to make biologists suspect that the cloning of a mammal was impossible. Thanks to these dazzling feats of reproductive science, and helped in part by his considerable presence, Illmensee was in great demand at conferences and on the international scientific lecture circuit. In a way he even helped create Dolly: his lectures would help to inspire two key figures in her story, Keith Campbell and Steen Willadsen. Keith saw Illmensee's lecture in 1984 at the University of Sussex, when he was beginning his doctorate. "Who knows, without him, I might not have got involved in this field at all."

But others had problems in reproducing the amazing feats of Hoppe and Illmensee, which have never been repeated. And although anecdotes may inspire and intrigue scientists, they don't sway them. Illmensee was secretive about his methods, and his reputation for having "golden hands" began to wear thin. Even members of his own group grew frustrated and mutinous. In 1983 matters came to a head when a commission of inquiry accused him of sloppiness and asked him to repeat a series of experiments. Crucially, however, these experiments were not concerned with cloning, and there was inadequate evidence that the pair had fabricated data.

Among the attempts to follow up this work, one stands out because of its impact on future thinking. More than anything, it explains why some scientists greeted Dolly's arrival with gasps of disbelief. Ironically, though reproducible (unlike the experiments that inspired them), this one study held back the field for years.

Well, that is at least according to many popular accounts of the history of cloning that describe an attempt to reproduce what Ill-mensee and Hoppe had done with a painstaking program of experiments at the Wistar Institute of Anatomy and Biology, in Philadelphia. The studies were carried out by Davor Solter, by then a respected embryologist and an American citizen, and his student James McGrath. In common with others, they wanted to learn more about differentiation. But, like everyone else (such as Chris Graham in Oxford), they could not get the Illmensee/Hoppe method to work. When Illmensee paid a visit to the Wistar, they asked him to give them a demonstration. For one reason or another, he never did. Perhaps Illmensee and Hoppe had been able to get three cloned mice but unable to repeat their good fortune. The efforts of McGrath and Solter, however, suggested otherwise. Solter himself believes that the "charitable explanation is that Illmensee and Hoppe mixed up some embryos and thought they had succeeded."

Their technique, and all the efforts to replicate it, were doomed to fail. Illmensee and Hoppe had carried out nuclear transfer into enucleated zygotes—fertilized eggs—rather than into eggs, because it was thought (wrongly) that the former would provide better support for development (though this baffled researchers who did the pioneering work with amphibians, where unfertilized eggs were used). "The original experiments have since been repeated many times by successful cloners and the results were always negative," said Solter. But that did not mean the exercise was a waste of time.

In attempting to clone mice this way, Solter and McGrath had made a useful contribution to nuclear transfer technique that would be described in *Science* in 1983 (after being rejected by the rival journal *Nature* as having little practical application). Until then scientists had not only penetrated the zona of the egg to carry out nuclear transfer but pricked the inner—plasma—membrane too. As Solter put it, "Everyone was killing eggs by jabbing needles into them." The Wistar team established a gentle way to remove the nucleus from a fertilized mouse egg and then introduce a new nucleus with as little disturbance as possible, leaving the inner membrane of the egg or zygote intact. We drew on this wisdom in the creation of Dolly. They created live mice by nuclear transfer this way (though not clones, since they could make only one copy of each zygote). "Weeks after our paper came out several laboratories started successfully using our technique in attempts to clone large farm animals," Solter recalled.

He and McGrath went on trying to reproduce Illmensee's work in a tenacious and methodical set of experiments. They had already shown that the simplest form of nuclear transfer—from one newly fertilized egg to another—succeeded, underlining how their technique was capable of delivering results. That was about as far as they could go, however. Almost all transfers from two-cell embryos reached the blastocyst stage, but transfers from four- and eight-cell embryos and from the inner cell mass cells of more advanced embryos (used by Illmensee and Hoppe) did not. They had no success with using eggs as recipients either.

Their failure had been so broad and so systematic that their results were published by the journal *Science* in December 1984. The Wistar results were more than negative; they effectively ended Illmensee's research career because of the questions they raised about how he obtained his results. Hoppe carried on trying to find the magic batch of cytochalasin to enable him to create uniparental mice, an impossible feat. The Wistar team concluded that "the cloning of mammals, by simple nuclear transfer, is biologically impossible." And that was that. Spemann's "fantastic experiment" remained a fantasy.

FROM THE ASHES

TODAY DAVOR SOLTER is the director of the prestigious Max Planck Institute of Immunobiology in Germany. His sandy hair is going gray, and he can sometimes look burdened. Despite many years on the international circuit, he still speaks English with a Croatian accent. His colleagues find him a warm, charming, soft-spoken man who is generous in discussing his findings and sharing his insights. He becomes animated when he discusses his passions—films and books (Philip Pullman is a favorite author)—but on certain scientific subjects, such as the history of cloning, he often becomes cross.

Solter is weary of the traditional depiction of his 1984 *Science* paper: as a hammerblow to efforts to clone mammals, one that dented the confidence of people like me and held back the field for years. The Wistar studies did indeed suggest that the potential of transferred nuclei to support development fell off a cliff face within the first few cell divisions of an embryo. Solter and McGrath suspected that it had something to do with how, compared with amphibians, the genes in mice go into action so early in development that they could not be "reprogrammed" by transfer to an enucleated zygote. In *Science* they suggested that the key to the problem might be found in earlier studies that they had conducted at the Wistar, as had Azim Surani of the Institute of Animal Physiology in Cambridge, England, in which mouse embryos that had two fathers or two mothers (created by exchanging maternal and paternal chromosomes) all died. The implication was that the embryos needed genes from each parent.

We now realize that with this work Solter and McGrath had made a key contribution. This phenomenon, called imprinting, ensures that, for certain genes, only the copies from the mother or father are used during development. This, too, would make reprogramming harder, so they concluded in *Science* that their results *suggest* that the cloning of mammals by simple nuclear transfer is impossible. Given what we know today, their thinking was far from wrong. But they used zygotes rather than eggs, an approach that we now know was less likely to be successful. The Wistar team was also unwise to use the word "impossible" and has paid for ending on this wrong note ever since.

But perhaps what rankles Solter most of all is his common portrayal as the powerful scientist whose devastating judgment held up the field of cloning. He protests that this is laughable, given his position in the field in 1984. "People were already doing nuclear transfer in farm animals with promising results for the past year and nobody that I spoke with at the time was in the least dismayed with our results." Several teams, including my own, were indeed pressing on, not least because there are hundreds of factors that affect nuclear transfer, and, though he had been diligent and thorough, Solter had not explored all the possibilities.

We were mere animal scientists who worked in fields, barnyards, and slaughterhouses. To us there was practical value in being able to make several copies of embryos recovered from the very best farm animals. Some registered elite animals have values of around $1 million each, and their offspring sell for up to $200,000. Potential profits provided the stimulus for us all to go on. Unknown to us at the time was the fact that nuclear transfer is easier to carry out in many farm animals than in that ubiquitous laboratory animal, the mouse. Researchers who were developing methods to clone farm animals included Neal First of the University of Wisconsin, where Randy Prather, Frank Barnes, Mark Westhusin, and James Robl, among others, worked on how to clone prize cattle embryos with the backing of the W. R. Grace Corporation.

Moreover, the gloom of Solter and McGrath did not put off Steen Willadsen, an inventive, extrovert, and energetic Dane whose contribution was recognized with the Pioneer Award of the International Embryo Transfer Society. I am sometimes called the "father" of Dolly (unfair, given the huge contribution of Keith Campbell and others at the Roslin—and somewhat daft too). Well, this brash and mysterious Dane is without doubt her "mother." Willadsen began as a veterinarian, then worked as a research scientist, first in Britain and later in the United States, to make many important contributions to the field. Now working on human reproduction, Willadsen is a brilliant maverick who does not suffer fools gladly.

Willadsen had also worked under Chris Polge at the Agricultural Research Council Unit of Reproductive Physiology and Biochemistry in Cambridge, where I had frozen boar sperm and produced the first calf from a frozen embryo. Just before he arrived, I had departed for the Roslin Institute. Willadsen inherited my fellowship and my research project in an environment of great academic freedom, where good ideas and innovation could flourish without interference. And, of course, he could thus work on an apparently discredited field such as mammalian cloning.

Willadsen played around with cloning by embryo splitting. One problem is that when the protective zona is damaged, the mother's system destroys the embryo up to the blastocyst stage. Willadsen overcame this by dividing a sheep or cow embryo in two, then encasing the halves in a couple of millimeters of protective agar, a jelly made from seaweed. That way, the embryos could be allowed to develop in the reproductive tract of another female before recovery—so the agar could be picked off with a hypodermic needle once it had done its job—and then implanted in a surrogate in which it was advanced enough to survive until term. As well as making it easier to clone by embryo splitting, this technique made it much easier to manipulate embryos, expanding the horizons of reproductive science. With Carole Fehilly, who would become his wife, Willadsen even mixed cells from embryos of different species to create the geep (goat/sheep).

This was not the first man-made chimera. That honor goes to a mouse born on 6 March 1961 as a result of an experiment by Andrzej Tarkowski, then in the Department of Zoology at University College of North Wales, where he worked as a Rockefeller Foundation Fellow. Almost four decades later, in the *International Journal of Developmental Biology*, the Polish embryologist cheerfully referred to his "crazy" project. He found that you could combine two mouse embryos consisting of a few cells and end up with a single animal whose salt-and-pepper fur was the only visible token of its dual origin.

At the same time, and to his great surprise, he discovered another scientist doing exactly the same crazy experiment—Beatrice Mintz of the Fox Chase Cancer Center, in Philadelphia. Animals made from two or more embryos were dubbed chimeras, after the ancient Greeks' famous fire-breathing amalgam. Like Driesch's flexible sea urchin embryos, chimeras underlined how the cells in an embryo responded to their environment. Another way to make them, by injecting cells into blastocysts, a method devised by Richard Gardner and Ralph Brinster, would open up even more possibilities for introducing foreign cells into the embryo.

Willadsen and Fehilly were the first to mix cells from different species. This apparently bizarre experiment to create a geep chimera had a serious purpose. At the time, they were interested in determining what makes a fetus tolerated by the mother during pregnancy, an immunological issue that sheds light on what can go wrong and lead to miscarriage. Although a sheep will not tolerate a goat embryo and vice versa, both will tolerate a geep embryo. Furthermore, some cells in a chimera embryo form the fetus, whereas others form the placenta. This explains another rationale for Willadsen's research: this approach could allow him to create a chimera in such a way that a common species could give birth to a fetus from an endangered species: you could make the chimera so that the placenta forms from one set of tissues (common species), and the fetal tissue from another (endangered).

Willadsen pressed on with his cloning research. In 1986 he stunned everyone by announcing in a paper in the journal *Nature* that he had cloned sheep from early embryos. This work is now recognized as a critical milestone in reproductive science. Because Illmensee's feats had not been verified and McGrath and Solter did not assess development to term, Willadsen's lambs were the first mammals of any kind to be cloned—beyond any doubt—by nuclear transfer. He had, of course, tried to achieve the feat with Illmensee's methods. But he had failed, like everyone else. Undeterred, he overcame the problems by sheer grit. When I once visited him in Canada, I was amazed that Willadsen did much of his research solo. He made up for the lack of helpers by working hard. Cloning requires many different skills, and usually they are provided by a big team, as will become clear in the next chapter. Yet in this laboratory he often did all of this work on his own, even down to the anesthesia and surgery on the same animal.

He also contributed boundless ingenuity. When Solter and McGrath tried cloning, they had used one-cell embryos as recipients for their donor nuclei. But Willadsen reasoned, rightly, that an unfertilized egg would be better because it was brimming with machinery—proteins—to act on the DNA from sperm and this machinery could reprogram the DNA in a donor nucleus. There was all the amphibian work to suggest this was the way to go.

Looking back to when he wrote the 1986 *Nature* paper on cloned sheep, Willadsen said,

> I made sure that my "predictions" were already realized by the time I let *Nature* publish them. By then I had not only cloned cow embryos, but also integrated embryo sexing and embryo freezing into the procedure, and I had demonstrated that serial nuclear transplantation worked in principle. In other words, I had established a framework, within which one could hope to tackle all reasonable cloning needs of cattle breeders—which was the official objective I had stated to the ARC.
>
> Before hardly anyone knew what was happening, a nuclear transplantation procedure of general usefulness had been developed (1983–84) and both sheep and cow embryos had been successfully and repeatably cloned whereby I mean that lambs were born (August 1984) and calves were under way (November 1985). And in so far as this could be done by one man, I did it.

Bravo! To achieve that as a member of a team is impressive. To do it alone, and create two lambs from three eggs, is heroic.

There was more. Working in a Texan ranch for Granada Genetics, a company that eventually fell by the wayside when a tax break was ended, Willadsen pushed back the boundaries once again. While some felt that it would not be possible to use visibly differentiated cells for nuclear transfer, Willadsen had other ideas: a pragmatist, he would obey this "law" of biology only after he had probed and tested it himself. All that bothered him was that it became harder to obtain these cells because they clung together more tightly in more advanced embryos.

Going against the grain of dogma, Willadsen had produced live calves by 1987 by nuclear transfer from embryos that had progressed to the 32- to 128-cell stage, but this work was not presented in public for several years. At this stage of development the cells of the embryo can still readily be separated. The nucleus from one of these cells was then transferred into an aged egg. Instinctively, one would have expected a fresh egg would be better, as is the case for *in vitro* fertilization. Yet the eggs he used were older than those that are used for IVF, for example, a fact that rather shook me at the time. With hindsight it is possible to see that Willadsen had, by trial and error, chanced upon one of the secrets of the Roslin team's later success. By using old eggs, he had found one of the effective ways of cocoordinating the growth of the donor cell nucleus and the recipient egg. (Later research revealed why this is important. An aged egg makes it easier to activate the development of the clone. In our later research, however, we discovered that old eggs do have one disadvantage— they are not so good at reprogramming a transferred nucleus and so would struggle to reprogram the DNA in an adult cell.)

Willadsen's success was remarkable. By this stage of development the early embryonic cells had differentiated into the first two cell types—the outer trophectoderm, the shell of cells that becomes ancillary equipment such as placenta, and the inner cell mass, which is destined to be the embryo proper. His research suggested for the first time that nuclear transfer in mammals was possible from at least these early embryo stages. Willadsen's work would provide much of the impetus for Dolly.

That same year of 1987 also saw Illmensee's controversial feat, the one that originally inspired Willadsen, at last come to fruition in a convincing way. Yukio Tsunoda of Kinki University and colleagues announced that they had cloned mice after transfer of nuclei between very early embryos. Illmensee claimed that this vindicated his work. Crucially, however, Tsunoda had used a different approach. As Solter drily remarked at the time, "The moon landing by the Apollo mission did not prove that the techniques and results described by Jules Verne actually worked."

WILLADSEN'S INSPIRATION

MY PATH TO DOLLY began with a stroke of serendipity while drinking with colleagues in a Dublin pub in January 1987. I was attending my favorite conference, the annual meeting of the International Embryo Transfer Society. In those forlorn days of relentless cutbacks by the British government, I was lucky to be there at all. I was the guest of my director, Roger Land, who had some soft money from a consultancy, which he used to fly me to the meeting. Despite our many differences, I owe Roger a great debt for giving me that opportunity. In the dimly lit pub I was told of pivotal and illuminating new work by Willadsen, now working in the heart of Texas, by one of his colleagues, the veterinarian Geoff Mahon (whom I thought of as "Willadsen's minder").

These informal chats are often the most important part of scientific meetings, when fellow researchers will gossip and sometimes share their ideas or new results that have not yet been published in formal papers. (Sadly, the exchange of ideas has diminished as research has become more commercial and there is more pressure to patent.)

From Geoff I learned that Willadsen had used nuclear transfer with a bovine embryo that was a few days old, at the early blastocyst stage, when they had around sixty-four cells that could still be separated. Of course, this experiment for Granada Genetics at College Station, Texas, could not possibly have worked, according to the dogma of the day, because the genes of the cattle embryo had already gone into action and its cells started to differentiate. But Geoff told me that Willadsen had succeeded. Various pennies dropped and bells clanged (I am not the sort of person who runs naked down the street shouting "Eureka!").

While Roger Land and I flew back over the Irish Sea on a small and rather noisy plane, I enthused about what I wanted to do. The very fact that Willadsen had produced calves from blastocysts almost certainly meant that he had carried out cloning with the cells of the inner cell mass, the part of the blastocyst that mostly turns into the embryo: working in Oxford, Rosa Beddington and Liz Robertson had already shown that mouse embryo stem cells retain many of the characteristics of the cells from the inner cell mass.

Now I could see how to make genetic modification easier because experience in the mouse was showing that precise genetic change was comparatively efficient in embryo stem cells. This meant that if we could derive embryo stem cells from livestock we would be able for the first time to make genetic change in the embryonic stem cells and, only when it worked properly, create a clone from one of these GM cells. Together the techniques of stem cell derivation and modification and nuclear transfer from them would revolutionize genetics in livestock. As well as being less hit and miss, this approach would not rely on so many embryos and be much quicker.

I managed to visit Willadsen, who was by then working in Calgary, Canada. He was not only fun and hospitable but also a gentleman (when scientists say someone is a gentleman, or generous, they mean he openly discusses his ideas and reveals those apparently insignificant details of experimental method that are crucial for success). As he carried his young child around with him in the scrubland near his laboratory, Willadsen gave me plenty of useful advice. Later he showed me the technical details and even gave me a demonstration of how to protect newly constructed embryos with a coat of agar, forming shells that vary from only a fraction of a millimeter to a couple of millimeters across.

But there was one problem. No one had yet isolated embryonic stem cells from sheep. Even as I write this, no one has pulled off this feat. As it turned out, this would not matter, because of a twist in this story, one that led to Dolly. With my colleagues I would discover that you could, amazingly, extend Willadsen's feat so that it was effective with adult cells, which supposedly were already committed to be a skin or blood cell or whatever. We could turn back the developmental clock. We could rejuvenate cells. We did not have to grow sheep embryonic stem cells in order to genetically alter sheep.

MEGAN AND MORAG

HERE IT IS WORTH standing back for a moment to emphasize two things about the work that led to Dolly. After years of involvement in pioneering projects to add genes to livestock, we were all too aware of the limitations of early techniques to genetically alter animals. They were inefficient and unreliable and, worst of all, gave no possibility of deleting harmful genes or implanting new ones in a precise way. Our hope, as we sought funding for the next stage of our research, was to achieve two new things: to derive embryo stem cells from livestock and to be able to transfer nuclei from them. If we could do that, we could carry out genetic surgery on those nuclei. Mouse geneticists were by then accustomed to being able to make, or "target," precise genetic changes through the use of embryo stem cells. This technique had revolutionized molecular genetics in the mouse. We wanted to create comparable opportunities in breeding farm animals to make them hardier, faster growing, and disease resistant.

Second, our achievement rested on the hard work of a lot of other people. The group at the Roslin is sometimes credited (wrongly) with the development of nuclear transfer in mammals. As is now clear, that credit is due mostly to Willadsen. And, of course, the origins of his work stretch back even further. Indeed, I did not even introduce the nuclear transfer technique to the Roslin Institute. Until that time I had been interested in embryo development in the womb and, in particular, why so many embryos perish even in healthy females and what the role of maternal hormones is. A Brazilian vet from a British family, Lawrence Smith, came to work for me at the Roslin to do experiments on the egg in the oviduct. After a few weeks he decided that my original suggestion was boring and wanted to try his hand at nuclear transfer instead. Wise choice. By then, of course, Willadsen was blazing the trail ahead.

It was Lawrence Smith who taught me the basics of cloning. Together we produced Roslin's first cloned lambs—four in all—by nuclear transfer in 1989. As well as help confirm Steen Willadsen's findings, that work marked the Roslin as a center for cloning. Lawrence also taught Bill Ritchie, who carried out the nuclear transfers in our later critical experiments that led to the birth of Dolly. As the media like to tell us, the principle of cloning is extremely simple. You need two cells in order to produce a clone. One is an egg recovered from a donor ewe at about the time when normally it would be mated. The other is an adult cell from the sheep you want to clone. First you remove the genetic information from the egg. Then you introduce genetic information from the adult cell by fusing the two together. This "reconstructed egg" is grown for a few hours or days in the laboratory before it is transferred into a surrogate mother and given the chance to develop to become a lamb. Five months after transfer of the embryo to a recipient ewe, one can expect to see a cloned lamb—if one is very lucky. Easy.

More details of the grind, slog, and sweat behind the feat later. Crucially, Lawrence had shown that one important ingredient of success was the state of the nucleus of the donor cell—that is, whether it had just divided or was preparing to divide. He completed an exquisite analysis of the effect of varying the combination of early and late stages in mouse nuclear transfer. This insight was extended when, in 1991, the Roslin's nuclear transfer effort was joined by a long-haired chap from Birmingham with an antiestablishment demeanor and, to complete the stereotype, a love of sinking pints in the pub. Our new "postdoctoral higher scientific officer" was Keith Campbell, who was to play a central role in the Dolly project.

The son of a seller of seeds, Keith would fill his mother's kitchen with frogs as a boy. He did not exactly shine at his grammar school, leaving without A levels. He had also been a medical technician ("I resigned my job the day I qualified to do it"), taken a job in a pathology lab in Yemen, returned to help fight Dutch elm disease in Sussex (which was devastating Britain's elms), and then worked on cancer. When we met him, Keith had an arty look about him and, as Colin Tudge, the coauthor of our book, *The Second Creation*, once put it, something of the mien of a folksinger.

As he talked with me and my daughter (Naomi was spending

time at the Roslin for "work experience") over lunch, we both were struck by the clarity of his ideas. Keith recalls it differently: "I ranted on for about an hour and a half, and they all fell asleep." The author Fay Weldon memorably described him as being as "wild-haired as a genius should be. Charismatic, inspiring, an ideas man come to this kind of genetic technology recently, a schoolboy so interested in tadpoles that his mother used to have to sweep baby frogs from the kitchen."

Keith had the ideal technical background for what I wanted: for his doctoral project in Sussex University, he had studied the cell cycle in *Xenopus* and was familiar with the work of Gurdon and Illmensee and, much more important, was fascinated by it. He had also picked up a profound insight into differentiation during his first doctorate, while working for a cancer research charity at the Marie Curie Institute in Surrey, an effort that petered out because his supervisor became ill.

Because of this background, he suspected that Spemann's fantastic vision could be realized and adult cells be used in nuclear transfer. After all, he reasoned, certain tumors consist of a blend of hair, bone, fat, muscle, and other cell types that probably originated from just one adult cell, or from transdifferentiation. Indeed, stem cells are now thought to play an important role in cancer because they replenish tissues and organs throughout a person's lifetime, and their progeny are more vulnerable to accumulating enough genetic changes to become cancerous.

The Roslin in turn offered Keith, who had a partner and one daughter at that time, a steady income, employment near where they lived (Keith was by then doing a postdoctoral fellowship at the University of Dundee), and a chance to clone and apply his ideas about the cell cycle and differentiation. Whereas Keith was driven by basic curiosity alone, I also wanted to use sheep embryonic stem cells for cloning. In this way I hoped to create new opportunities for genetic modification: modify the stem cell and, only when you have it absolutely right, go on to clone and to implant embryos.

Keith focused on the process of nuclear transfer itself. His first critical contribution to the cloning effort at the Roslin was to investigate what it is about the state of the nucleus that primes it for nuclear transfer. He found that the efficiency of nuclear transfer was very limited—unless, that is, the cycles of the donor cell and egg were coordinated.

Here it helps to think of a cell as a very exotic chemical clock. Each beat—or division—of that clock is controlled by a complex cycle of chemical reactions, and, unsurprisingly, to make cloning work well, you have to understand this mechanism. If cloning is attempted on the wrong "beat," as was done so many times before with more differentiated cells, it does not work at all. Indeed, the better we understand the baroque cycle that ticks within our cells millions of times each second, the more amazing the day-to-day business of life itself seems. And the early crude attempts at cloning by Spemann, Briggs, and others, though of limited success, seem little short of miraculous.

Thanks to the important work of Keith, using cow fibroblasts, we would find that for nuclear transfer to succeed, we had to use various ways to coordinate—strictly—the cell cycles of the egg and donor cell. One approach was to mimic Willadsen's method of using aged eggs. In our case we specifically stimulated the egg to activate before transfer of the nucleus. Such a preactivated egg is a universal recipient, because it is suitable for a nucleus at any stage of the cell cycle. However, as we now know, the fact that it is aged or activated means that the egg is not best able to reprogram the transferred nucleus, the crucial step that makes it possible to rejuvenate differentiated cells.

Alternatively, and preferably, the egg could be in a particular state, the so-called MII. Such an egg appears to have the best combination of factors for reprogramming. Appropriately enough, this is the same state that eggs reach when they mature in an ovary ready for fertilization. MII is shorthand for metaphase II, one of the stages in egg development (meiosis). The "II" refers to our being on the second cell division in meiosis. Metaphase is the stage in the cellular clock mechanism where the cell has no nucleus and the chromosomes have "condensed": they lie down its middle, ready to be duplicated and divided. The egg is rich, too, in a substance called meiosis-promoting factor (MPF; the M can also stand for maturation or mitosis, depending on which scientist you are talking to), which, as the name suggests, speeds the process along. We were lucky to use sheep, which meant that we could catch eggs in this receptive stage: rats race through metaphase II, and it is only by the use of drugs to brake their development that they have been successfully cloned.

The next stage is to prepare the MII egg for reprogramming the transferred DNA by emptying the egg of its nucleus, and thus its DNA. Because these egg cells do not have a nucleus, but rather have free chromosomes, the term "enucleation" is used loosely for the removal of the nuclear DNA. Crucially, because the nuclear membrane has melted away at this stage, special factors are present, which can bind to DNA to get the process of replication under way.

Just as important as the choice of egg is the control of the cycle in the cell that is to donate DNA to the emptied egg. The cell must be at the beginning of the cell cyle, ready to move on by copying the DNA and growing. This much we had established within a year or two of starting the project. A practical limitation to the use of this new knowledge was that embryo cells cannot be stopped at the beginning of the cycle. Our Spanish student, Pedro Otaegui, who was doing parallel experiments in mice, spent hours setting up embryos so that they were at an appropriate stage just when he needed them—working night and day, as is often the lot of students.

A colleague, Jim McWhir, had continued a quest for embryonic stem cells in sheep and other livestock that he had started with one of the pioneers of the field, Martin Evans, in Cambridge. Despite intensive effort he just could not keep the stem cells growing for more than a few passages (through a few cultures). Cells continued to grow, but they were not embryo stem cells. We agreed that it would be helpful to know when the cultured cells began to lose the flexibility (totipotency) as they replicated. We could do this by using Keith's methods of nuclear transfer to assess the potential of the cells.

So Keith, Jim, and I embarked on experiments to clone from cultured cells that had been through more and more passages, seeking that cutoff point when totipotency evaporated. Using the then standard method—we predicted—we would obtain lambs from early cultures, but not the later ones, showing that a change had occurred. Success in some of our cloning experiments led Jim to believe he had found the elusive embryonic stem cells (he called them TNT4 cells, totipotent for nuclear transfer). Today we know he was mistaken.

However, what we had was itself very interesting. These were cells that had only just begun to differentiate, and we found that we could stop them at the beginning of the cycle—we starved the donor cell to send it into a resting, or quiescent, state. Simply by our removing a source of nutrients for five days, the cell stopped at the point that we needed.

With this understanding, we cloned Welsh Mountain ewes, from the embryo-derived cells that had been cultured in the laboratory. We made each by fusing an embryo cell with an empty sheep egg using a spark of electricity. In this way Keith and Bill Ritchie produced 244

The Roslin Institute nuclear transfer group, including Keith Campbell at the back right, Bill Ritchie back left, and Lorraine Young front left. Jaki Young was the secretary working for the group at that time and is in the middle of the back row. (Photograph courtesy of the Roslin Institute.)

embryos, of which 34 developed to the point where they could be implanted. This was not many embryos, but we lacked the money to do any more. Then we held our breath. We knew that if these experiments worked, it meant that we had made embryos from clearly differentiated cells and we would have taken the science of cloning to a new level.

Originally five clones were produced, but three died—two within minutes of birth and the third within ten days, of a hole in the heart. The survivors were born a few days apart to different surrogate mothers in June 1995. The names of the lambs were 5LL2 and 5LL5, where the first 5 refers to 1995, LL is the farm designation, and the second number refers to the number of clones born that year.

The names of these pioneering sheep were changed by Marjorie Ritchie, the Scottish manager of our Large Animal Unit, who had helped bring them into the world. Marjorie, the veteran of hundreds of embryo transfer experiments, loved the camaraderie of traditional nuclear transfer (as she put it, they were "happy days but also very busy days") and married my colleague Bill, the cellular surgeon. She named the new arrivals M, after her own initial letter, Megan to honor the Welsh Mountain ewes that supplied their genes and Morag to celebrate the Scottish Blackfaces, the ubiquitous local breed that provided the eggs. We did not celebrate their birth, since they were merely two of a series of experimental animals, and, as Marjorie points out, it is only now that we appreciate how significant they were. They did make headlines when we unveiled them in 1996, but generated nothing like the fuss that was to come with Dolly.

Megan and Morag were very important to me. This was our first groundbreaking experiment. The embryo cells that had been used to create them with nuclear transfer were multiplied in culture until they were nine days old and thus partly differentiated. They carried molecular markers of differentiation too. These cells were from embryos significantly later than the blastocysts used by Willadsen in his critical experiment. The experiment was significant because it showed that, by paying attention to the details of the cell cycle, one could reverse this differentiation and "rejuvenate" cells.

As if to underline this, Keith became involved in almost every stage of the process. He moved out of the lab to shave the bellies of the sheep and disinfected them for operations, helped gather eggs and embryos from donors, and even slept in an office next to the sheep toward the end of their gestation so that he could regularly check for the first signs of birthing. He had been tailed by the local police as he went for a sandwich in the middle of one night ("they were just as bored as I was"). During the perpetual twilight of the Scottish summer nights, he shared this chore with Bill Ritchie, John Bracken, and others, before rushing home to say good-bye to his children on the way to school. As it turned out, all the lambs were born in the afternoon.

Megan and Morag proved that cloning from a cultured cell, one that was partially differentiated, was possible. They form one great clone with the other lambs born alive as a result of the process, plus the fetuses that did not make it to term, the embryos that never made it to implantation, and the cells used for nuclear transfer. Keith and I applied for patents on aspects of the technique in August of that year. Soon after we had begun to use starved cells, Keith recognized that it might have other benefits apart from the convenience of being easy to stop the cells' development. Because of this we did not talk to anyone about the work. As a result, after the lambs were born, we were able to submit a patent application. In the UK once an idea has been mentioned in public, it cannot be patented; in the United States, by contrast, there is the "year of grace" during which patent applications can be submitted even after a public announcement. This later enabled the Roslin to draw in millions of dollars to start a company and support institute research. For some reason that we do not understand, it seems that in this quiescent state it is easier to change the functioning of the chromosomes: they are amenable to being reprogrammed because this represents a necessary step on the path of cell differentiation. The DNA carries the instructions for the cell, but other mechanisms regulate the way in which they function— epigenetic mechanisms. These determine that a particular cell will function as muscle or brain or gut. During nuclear transfer these have to be reset so that the chromosomes are able to support development of an embryo.

This result was so important to us and to others in the field that, after the birth of Megan and Morag, we were confident that one day it would be possible to clone an adult and began to draw up details of just how to do this. In science the ability to plan this way is a tremendous advantage, and it confirmed that our success rested on a deep understanding of the cell cycle obtained by Keith.

The ability to use more differentiated cells for nuclear transfer gave us opportunities to carry out genetic surgery on the cultured cells before cloning. And thus they showed the way to create GM animals more efficiently. They also inspired other groups to investigate nuclear transfer. The next step—to be achieved within two years, I predicted at the time—was to "design" a sheep by adding new genes to a cell before fusion, guaranteeing desirable characteristics, such as an ability to grow quickly or to produce human proteins in animal milk that could be refined and used as drugs. All we would have to do was find one GM cell taken in a lab-grown line of millions and combine it with an empty unfertilized egg that could be harvested from slaughterhouse sheep.

Our experience with Megan and Morag also showed that reproductive cloning was risky. Of 244 attempts they were the only survivors. We wondered whether we had been lucky or unlucky—whether the real efficiency was much worse or much better—and vowed to repeat the experiments to find out. Both clones were later offered the chance to mate, though only Morag seemed to take up the offer and went on to have lambs. Alas, Morag succumbed to the same lung disease that afflicted Dolly, though Megan has lived on to a ripe old age of ten and counting. Davor Solter wrote an accompanying editorial in which he stated that "cloning mammals from adult cells will be considerably harder, but can no longer be considered impossible." The media were not slow to pick up on the implications (except, notably and surprisingly, in America). As the *Daily Telegraph* reported in Britain, "In theory, it is now feasible to clone human offspring without the participation of a man." Lord Winston, one of the country's leading figures in

The celebration when Megan was ten years old, in July 2005. She is seen with Ian Wilmut, Marjorie Ritchie, and John Bracken, who were involved in that experiment, and Tim King, who is now the veterinary surgeon in charge of the facility. (Photograph courtesy of Roger Highfield.)

reproductive science, was at pains to point out, "I don't think the human male is redundant. I think the human male is unnecessary."

With the help of colleagues at neighboring PPL Therapeutics, we cloned Dolly in 1996 and unequivocally realized Spemann's dream of decades earlier. The streamlined accounts that appeared in newspapers do little justice to how we restored to adult cells the full potency and potential of the embryo. Just dwell for a moment on what we were aiming to do. We were trying what, for most people, was unthinkable: to achieve a live birth without the help of a father. That meant carrying out complex surgery on an egg that is so small that you can only just see it with the naked eye. Whether whale, mouse, sheep, or human, the eggs used for cloning mammals are approximately a tenth of a millimeter in diameter, around one two-hundredth of an inch. Now imagine trying to turn two cells—the egg and another cell—into one. Then we had to persuade the resulting reconstituted embryo to divide successfully. And we had to some-how reprogram the cell so that its DNA did not "remember" ever having been an adult cell but was sent back in cellular time to the embryonic stage.

We operated on sheep eggs, using an astonishingly blunt instru-ment—a microscopic pipette with an inside diameter about the same size as the egg itself. We removed the egg's DNA, a molecule meas-uring 2.5-billionths of a meter across, and replaced it with the usual genetic complement of a cell, 2.3 meters or so of DNA—about eight picograms in all; that is, eight-trillionths of a gram. Ultimately, the effects of this surgery would have consequences at the molecular level: the egg is packed with thousands of complex structures such as mitochondria and millions upon millions of molecules.

Astonishingly, we found a way to fool the egg by a shock of elec-tricity into thinking it was a developing embryo. Within each cell of a normal embryo, complex mechanisms normally ensure that genes are turned on and off at the right time in active regions of the nucleus, a membrane-bound region at the heart of the cell, to lay down tis-sues, organs, and other structures. To succeed, we thus had to find a way to persuade the egg to turn on the genes in the newly introduced DNA that are usually active in an embryo, turning off those that work only in adult cells, and all without disrupting the careful cho-reography of genes that subsequently blink on and off during years of development. Put this way, the hurdles sound insuperable. Yet we succeeded, because of our finding that starving the donor cells was crucial if the genetic choreography within the reconstructed embryo was to have a chance of developing normally. We managed—albeit imperfectly—to reprogram an egg with adult DNA when we really had only the vaguest idea of the molecular details of the miracle of development. An "impossible" lamb was born. This is how we made Dolly.

THREE

DOLLYMANIA

SQUASH BALLS, infinite patience, a cramped, warm room, and udder cells from a long dead sheep are just some of the ingredients of animal 6LL3. With the help of an egg, glass tubes little thicker than human hairs, and a flock of ewes, a big team at the Roslin derived every one of Dolly's cells, from her nose to her hooves, from a single udder cell. We still don't understand the nitty-gritty of the cloning process, but, despite these gaps in our knowledge, the implications of this feat for science, medicine, and society are huge.

One of the twists of Dolly's story is that amid the media hubbub that surrounded her birth, and the acres of newsprint discussing the possibilities of cloning humans, only the sketchiest details of how we cloned her and the reasons we did it were discussed. Few commentators seemed to realize that we did not set out to create Dolly at all, at least not as a primary objective when we started out in 1995. She was actually a detour in our efforts to genetically alter farmyard animals to make new breeds that are healthier and more productive.

As we began to prepare our next steps after the births of Megan and Morag, several ideas were in the air. Could we use Keith Campbell's new step of inducing quiescence to clone from even more differentiated cells—from fetuses or adults? There was also obvious commercial interest in the idea of adding genes. PPL Therapeutics, a biotechnology company sited a few hundred yards away from the Roslin, wished to add genes to sheep cells while they were cultured. Perhaps with cells from embryos of the kind that we had used to produce Megan and Morag? Significantly, as it would turn out, PPL also had to hand a line of cultured, *adult* sheep cells, something that my group lacked, for its studies of how to persuade sheep to produce drugs in their milk. We assembled at the Roslin Institute in the autumn of 1995 to study the possibilities.

Alongside me in the group that managed the project were Keith, our long-haired cell cyclist, and Jim McWhir, the Roslin's stem cell expert. From our partner, PPL, there was Alan Colman, PPL's research director, who had done his doctorate with John Gurdon, the pioneer who had failed to clone adult frogs from adult cell nuclei (Alan later confessed he would have vetoed our work with adult cells if there had been embryonic alternatives: when Dolly was born, Gurdon himself said he was shocked). Also from PPL came Alex Kind, who would go on to combine cloning with precise genetic alteration, and his wife, Angelika Schnieke, a Ph.D. student, who at a critical stage in the sheep season had the good idea—backed with persuasive arguments—of using in our experiments mammary cells that had been collected from a pregnant ewe.

Many others at the Roslin were involved in Dolly's creation, such as our cell surgeons, Karen Mycock, a twenty-eight-year-old who had already cloned cows for a company called Grenada Biosciences in Newcastle, and Bill Ritchie, a neat man of military aspect who likes to go "Munro-bagging" (climb the 284 Scottish hills that ascend over three thousand feet, a list first compiled by Sir Hugh Munro). In addition, there was Bill's wife, Marjorie, who organized the surgery, operated on the sheep, and provided a dependable supply of banter to keep the mood convivial and relaxed in the operating theater.

Our multipliers of cells were Patricia Ferrier, Ray Ansell, and Angela Scott. Mike Malcolm-Smith helped with surgery, and Douglas McGavin and Harry Bowran injected sheep. And there was John Bracken, a very experienced stockman with the quiet demeanor and that "firm but gentle" manner that all animal handlers must cultivate, who coined the name Dolly immediately after she was born. Standing in the pen, John said to Douglas McGavin, "You know what we are going to have to call her, don't you? In view of the choice of cells for the experiment there is only one possible name: Dolly."

The UK government's Ministry of Agriculture, Fisheries, and Food (MAFF) had agreed to fund our work with fetal and adult cells after we had succeeded in using the more immature cells. In the application for the money, submitted two months after the births of Megan and Morag, I had stated that the cost of an experiment to clone embryonic, fetal, and adult cells would be an implausibly precise £484,519. A helpful civil servant played a key role at this point in Dolly's story. The project was funded only because of the strong personal support from John Caygill, the plant scientist who was our project officer in MAFF.

We got off to a good start in the autumn of 1995 and transferred nuclei from fetal cells and one group of embryo cells. But PPL did not produce as many cultured embryo cells as planned. It thus happened that early in 1996 we had the time, sheep, and money to do other things: we decided to work on adult cells. And so started the effort that would culminate with Dolly.

THE MANIPULATORS

THE CREATION OF DOLLY entailed surgery on an egg much smaller than the tip of a sewing needle. To carry out this microscopic surgery, we needed people with steady hands, good coordination, persistence, and patience. That ruled me out. Over the years, however, Roslin had trained a fair few "manipulators" to do such fine work. Bill Ritchie and Karen Mycock, who did the cell surgery in the

Dolly project, have the unique skill of being able to concentrate for hours as they stare into a binocular eyepiece, yet remain relaxed enough not to make mistakes.

Some manipulators enjoyed music as they worked on embryos: while pursuing his doctorate, Pedro Otaegui would listen to Leonard Cohen's downbeat songs of longing and despair. Bill and Karen would work side by side in a tiny room—a glorified cupboard—on the ground floor of the Roslin, with one doing the cell surgery as the other fused cells to create the embryos. "We used to sit and talk about a lot of rubbish really, or gossip, to pass the time," said Karen. "We could be in there for hours. It was like being in an isolation tank." Since the work was done during the short winter days, "sometimes you did not see daylight. You went into the lab during darkness and left during darkness."

The "micro" part of the micromanipulator rig that Bill used refers to how the cells that were to be combined by cloning were magnified with the help of lenses and optics. The microscope stood

Bill Ritchie seen at his microscope with manipulator equipment. (Photograph courtesy of the Roslin Institute.)

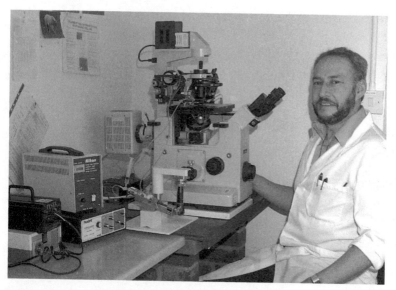

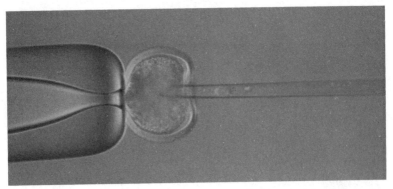

Partway through the process of enucleation the large pipette is holding the egg in place while the other passes through the zona pellucida and sucks out an area of tissue where the chromosomes are believed to be. (Photograph courtesy of the Roslin Institute.)

on a desk that was, in turn, mounted on a heavy metal plate, which, in turn, sat upon squash balls to absorb the smallest vibrations. Without a way to soak up unwanted movements, a passing lawnmower outside the laboratory window, the slamming of a nearby door, or racket from a radio would have made his cellular surgery shake, rattle, and roll.

The "manipulator" part of the instrument's name refers to the micromanipulator's two arms, each of which brandishes a fragile, hairlike pipette, through which oil can be gently sucked or blown to pick up a cell, and put it down. On one side is a pipette one-tenth of a millimeter in diameter, which is almost the same diameter as the egg that it is used to grasp. On the other is a much smaller pipette, which can be as small as one-thousandth of a millimeter and which is used to remove DNA and to move the donor cell about.

Bill moved the pipettes by twiddling two knobs (more modern versions use a joystick like those seen on computer games, only more precise). To pick up the egg, you apply suction with a syringe. To set it down again, you release the pressure by pushing in the plunger of the syringe. Micromanipulation is as fiddly as it sounds, requiring patience and months of practice. A high degree of coordination is essential, and the work is extremely repetitive, leading Bill to liken the requisite concentration to that required to fly a helicopter. "Three hours and you are knackered."

THE EGG HUNT

FOR CLONING, you first need a reliable supply of eggs—hundreds, if possible—because nuclear transfer is so hit and miss. Although females carry a lifetime's supply, they must be at the right stage of development. In a mother, whether sheep or human, eggs are ripened by a cycle of hormones, gradually depleting her supply over a lifetime. After birth her ovaries contain many thousands of eggs, each wrapped in layers of cells, in a structure known as the follicle. During the weeks leading up to ovulation, the egg and its surrounding follicle are matured by changes in hormones. In the days immediately before ovulation, the egg makes proteins, RNA, and other molecules that are to direct development just after fertilization.

Today we can take a slaughterhouse sheep ovary and ripen the eggs that it contains in the laboratory. But in the early days of our research, from 1985 until 1995, we had to work much harder and obtain ripe eggs at the source, by carrying out surgery on a ewe at a precise moment. Because sheep give birth in the spring, when their lambs are most likely to survive, and because the gestation time is around five months, Dolly's story begins in midwinter, when our technicians and shepherds rounded up the ewes, a cold, wet, muddy, hard, and smelly job. Each day they had just six hours of mean midwinter light to work by.

We used around sixty donor ewes to obtain the eggs, carefully orchestrating their hormonal cycles. Each ewe was first "sponged"; that is, a sponge impregnated with the hormone progestogen would be put in the vagina to keep the ewes from ovulating. We then injected another hormone, gonadotrophin, twice a day for four days to make the ewes produce more eggs than the natural number of two or three. A final jab would make the animal ovulate at the appropriate time, within a six-hour window, so we could be ready to remove the eggs.

In the Large Animal Unit in the nearby Dryden Farm, two of our team would clean the sheep, shave them, disinfect their pink bellies, and anesthetize them. Unlike cattle, which have to be moved with a hoist, a sheep is light enough to be lifted onto the operating table. Another two of the team would begin the effort to get the eggs, one operating and one assisting by passing instruments and so on. A fifth member would hunt for the eggs and pick them out of sterile fluids washed through the reproductive tract. Sometimes my team was even joined by a sixth member. I enjoyed the camaraderie of this effort, which went on two or three days every week over a period of several weeks. I did not enjoy the bureaucracy. A lot of paperwork was required to obtain the funds from several different agencies, report on past and ongoing projects, and satisfy the Home Office that we had taken care of the sheep.

Bill and Karen had to use eggs at just the right stage of development in order for nuclear transfer to have the best chance of success. When an egg ripens in a ewe's body, it must undergo meiosis, a strange form of cell division that carries out vital bookkeeping when it comes to the numbers of chromosomes that the eggs (and sperm) carry. We would collect the eggs when the ewe was in heat, when we knew the eggs were partway through meiosis. Once again, the details are more complex than many realize, and readers of a nervous disposition may want to jump to the next section.

The vast majority of body cells in a sheep package their DNA into twenty-seven pairs of chromosomes (we have twenty-three pairs in our cells). They are called diploid cells and multiply by a process of cell division that is called mitosis. But for an egg and sperm to combine to get a new life going, they have to bring half that number of chromosomes along to create the usual quota of chromosomes in an offspring. Sheep eggs and sperm start off with a normal number of chromosomes and then, by the process of meiosis, throw out half this number; in the case of the sheep, they are thus left with twenty-seven chromosomes and become "haploid" cells.

Meiosis can be thought of as a variation on the theme of normal cell division (mitosis). Two meiotic cell divisions of a diploid germ cell make four haploid sperm. But meiosis is different in eggs: half the chromosomes are ejected in what is called a polar body—in fact, a tiny cell that lacks enough cytoplasm to function. Thus, while the cell division in mitosis creates two viable daughter cells, meiosis in eggs creates only one. To further appreciate how complicated the details are, note that scientists divide meiosis into several stages: there are two cell divisions called meiosis I and meiosis II, which cull chromosomes, each of which is divided into further phases to describe different chromosome "dances" that take place within the egg—prophase, metaphase, anaphase, and telophase.

We discovered that one stage of meiosis works best for cloning: the second metaphase, just before the egg is shed from the ovary. At this stage, rather than being uncoiled in the nucleus, as it is when its instructions are being read, the egg's DNA is conveniently packed in short chromosomes that are assembled on a cellular structure called a spindle, ready for the next phase of cell division when some of the chromosomes are shed into a second polar body after the sperm penetrates.

Once the precious eggs were collected at the Large Animal Unit, they had to be taken back to the Roslin Institute so that cloning could begin in earnest. They were placed in a standard little plastic test tube with a cap, called an Eppendorf. Bill would pop the Eppendorf into his shirt pocket; Karen, lacking the pocket, would tuck the container of eggs under her bra strap to keep them warm. "You did not want them to chill," she said.

BREAKING EGGS

TO MAKE DOLLY required teamwork of military precision during those chilly winter days. Typically we would give the ewes the last injection on Monday and recover the eggs on Tuesday morning. That afternoon Bill and Karen would get to work on them. On the same day we also had to remove the fruits of the preceding week's work—cloned embryos—from sheep that had acted as temporary recipients to allow the embryos to develop. The interlinked cycles of work on the sheep that turned at the Roslin somehow reminded me of the vine that winds itself around one of the pillars in the nearby Rosslyn Chapel. For Christians it was the tree of life and for the Norse the tree of knowledge, Yggdrasil, on which Odin sacrificed himself to gain the secrets of creation.

Bill has remembered how Dolly was made during the last few weeks of the season and how it was touch and go whether she would be made at all. ("There was talk of doing it the next season.") Karen has recalled that, on the day Dolly was created, there were not many eggs to work with. "It was quite a slow day." For this little act of creation, the eggs had to be kept happy, which meant keeping them in the right conditions. The cramped room in which the team worked was kept warm. The eggs, like the cells used to donate DNA to the reconstructed embryo, were kept in a carefully prepared culture medium to mimic the chemical balance in the ewe's body. Typically, this medium contains salts, energy sources such as sugar, proteins, and amino acids. We also added antibiotics to stop bacteria from setting up home in the manipulator and ruining everything. To keep the medium slightly alkaline, as it is in the body (pH around 7.4, similar to that in a swimming pool), we used a chemical "buffer." Bill would

The chamber on the microscope in which the eggs are held for manipulation. A pipette can be seen at the right. (Photograph courtesy of the Roslin Institute.)

load this medium into a special chamber—based on microscope slides—and, to keep it from evaporating, seal it with a slick of mineral oil. The pipettes were passed through the oil to carry out cellular surgery.

Once he had placed the eggs into the micromanipulator, nuclear transfer could begin in earnest. To remove the DNA from (enucleate) the egg, Bill had to break through its protective shell, which is given the evocative label of "zona pellucida." This shell is crystal clear, as the name implies, though not brittle like a hen's egg. Nature designed the zona as a gateway that could be penetrated by one sperm, but no more. The zona has a coating that allows sperm to stick on, and penetration is achieved with a great deal of tail thrashing along with a cocktail of zona-dissolving enzymes in the sperm head.

However, the elasticity and toughness of the zona can vary among species and make a big difference to the nuclear transfer process. In the case of Dolly we penetrated the zona with pipettes with sharp tips. Sometimes we used a stepwise jab by exploiting the piezoelectric effect, whereby certain crystals can be made to distort when a voltage is applied to create imperceptible movements. The latter method was developed in Hawaii by Ryuzo Yanagimachi's laboratory to pierce the zona pellucida of the mouse egg, which is unusually elastic and hard to penetrate, compared with that of a sheep egg.

Once Bill was inside the zona, he had to find and remove the egg's DNA, which was coiled up in the chromosomes. At this stage of egg maturation, the chromosomes were grouped near the surface of the egg. Even though the chromosomes could not themselves be seen, Bill could get some idea of where they lay because they were packaged together on the spindle (the structure that helps separate chromosomes during cell division). However, they were still far from obvious. Sheep, like other farm animals, have dark gray eggs brimming with fatty droplets that obscured Bill's view.

In this fatty fog Bill had to play hunt the spindle. One way was to cheat by adopting a cunning strategy developed in the mid-1980s by Steen Willadsen: slice the eggs in half. Because the DNA is in a lump, it ends up in one half of the egg. Unsurprisingly, this radical solution cuts the size of the reconstructed embryo and its precious cytoplasm that helps the whole process of cloning along. It thus reduces the chance of achieving a pregnancy.

Another way to play the spindle game emerged a few years later. Yukio Tsunoda of Kinki University used special dyes that bind to DNA and glow blue-green in ultraviolet light. If the eggs were incubated with one of these dyes before being exposed to ultraviolet light, it was possible to see the chromosomes clearly, even when they lurked in the cloudy eggs of farm animals. Once again, however, there was a catch: ultraviolet light damages the eggs and makes them less viable, the way it wrinkles and degrades the skin of sun worshippers.

At that time the Roslin's own Lawrence Smith came forward with a way, which we still use today, to track down the chromosomes. He found that he could spot them by looking for small clearings in the cloudy cytoplasm. Because the DNA is clumped in the spindle, the surrounding fog of fat lifts a little in this one spot. This method works in about 70 percent of the cases. To be sure he had removed them, Bill used Tsunoda's DNA label and then screened the contents of the pipette he used to suck out the egg's chromosomes: if it glowed under ultraviolet light, then we had been successful in

To confirm enucleation the pipette is held in UV light under which DNA lights up to reveal the chromosomes from the egg and the second polar body. This enucleation was successful. (Photograph courtesy of the Roslin Institute.)

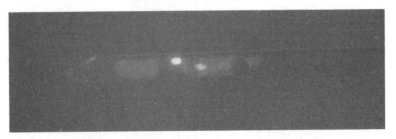

removing the labeled DNA. And in this way we did not upset the egg's cellular machinery by cooking it with ultraviolet light. (Jean-Paul Renard and his colleagues in France have developed yet another approach, in which they use extremely sensitive cameras attached to their microscope, enabling them to dim the ultraviolet light to such a low level that it could not do damage.)

With experience and patience, and Lawrence's method, an experienced manipulator like Bill Ritchie could remove DNA from up to one hundred eggs over a two-hour period, so long as the eggs were of good quality. They had a smooth outline, a uniformly cloudy cytoplasm, and were free of any clingy cumulus cells, the cells that nurture eggs during their development. (This measure of quality is, of course, partly subjective and one of many examples of the "witchcraft" of nuclear transfer.) Crucially, however, Bill did not remove all the contents of the egg: left inside was a jelly—the cytoplasm—that contains all the important chemical machinery to reprogram any newly introduced DNA. Now we had to obtain that DNA so that we could reprogram the empty egg.

THE UDDER DONOR

WE ARE NOW halfway through the process of nuclear transfer. The "enucleated" egg could be thought of a CD player that had just had its CD removed. To get the egg to play a new genetic tune, Bill required the equivalent of another CD, a new set of chromosomes. In the case of Dolly they were obtained from a cell in the mammary gland of an adult ewe. That is why we named her after the famously endowed Dolly Parton. None of us at the time could think of a more impressive set of mammary glands. The singer's agent later said she was honored by the tribute, joking, "There's no such thing as baa-aa-aad publicity." Her name was a gift for the headline writers: "Cloned Sheep's Name Brings Back Mammaries," said one. Another, which accompanied Dolly's announcement in the scientific literature, was entitled "An Udder Way of Making Lambs."

The mammary cells had originally been obtained during collaboration between PPL Therapeutics and Dr. Colin Wilde at the Hannah Dairy Research Institute, near Ayr, Scotland, to study lactation. At the time when the tissue was removed by surgery, the six-year-old Finn-Dorset ewe was partway through pregnancy and under general anesthetic. Years later, and long after her death, her cells were conveniently at hand for our cloning experiments. That is why there was no original donor animal to which we could compare Dolly; in retrospect, this would have been a useful thing to have been able to do. We came to grow weary of being asked about the fate of Dolly's identical sister/genetic mother.

Her mammary tissue was being used in research to increase milk production. One idea was to study modified mammary cells grown in the lab and select the ones where more protein was made, which was thought doable, then clone the best animal from it, which seemed unlikely at the time. As it turned out, Keith has recalled, the impossible part of this project worked and the plausible part did not.

We grew the donor udder cells in a culture medium that contained blood serum from fetal calves, chosen for the simple reason that it worked. That way we ended up with a large number of cells that were identical to the donors. We did not use all of them for nuclear transfer: some could be frozen and stored, almost indefinitely, in liquid nitrogen, which has a temperature of minus 196 degrees Celsius and, better still for cash-strapped researchers, is cheaper than beer. This way we could make clones years after the cells were taken, and long after the donor died. We could make another Dolly today from that same groups of cells, though she could never fill the giant hoof steps left by her older identical sister. Nor would we be allowed to do such a thing without a good scientific reason.

A healthy donor cell boosts the chance of success of the cloning procedure. For Dolly we used one simple health check, in which we counted the cell's chromosomes. This is one reason we cultured cells for nuclear transfer. As they grew in the laboratory, we could from time to time tot up the chromosomes in a sample of one hundred cells. If we found that a high proportion of cells had abnormal numbers of chromosomes, we threw the lot away. Critical to the success of our attempt to clone Dolly was the preparation of these donor cells before nuclear transfer. That meant using the cells at the right point in their cycle.

During each "cell cycle" there are precise mechanisms, still not fully understood, that ensure that during each period of cell growth the genetic information in the cells is not only copied but copied just once, and with great fidelity, in order for cell division to succeed. In this way the daughter cells contain exactly the same DNA as the parent cells. As was mentioned above, Keith Campbell was one of the first to emphasize the importance during nuclear transfer of ensuring that the cell cycle of the donor cell is coordinated with the cycle of the recipient egg. In effect, these cellular clocks must tick on the same beat. Usually this coordination is achieved by the use of cells at the beginning of the cycle, but this is difficult to arrange with embryo cells. The cells that Jim McWhir grew from the embryos had an advantage that they would stop at the beginning.

We starved the mammary cells of serum for five days, and they stopped growing and dividing to enter the state known variously as quiescence or G0. Intriguingly, this state had already been linked with differentiation, the process in which an embryonic cell evolves into a skin cell or whatever. This in effect gives the cell a breathing space so that DNA can be reprogrammed to determine its fate. The state can also be thought of as a form of hibernation, and it happens slowly and smoothly: whereas an electric motor simply stops at the flick of a switch, a cell closes itself down in a gradual, orderly way when starved. The cells no longer copy their genetic information and become comparatively inactive. We think that it is because the organization of nuclei in these quiescent cells is special in some way that they are more receptive to nuclear transfer than if they were in other stages of the cycle of cell growth and division.

The reason that G0 had been neglected in prior work was that cells from early embryos will not go into this stage of their own accord. Indeed, if you starve cells from early embryos, they do not hibernate—they die. Since G0 is a form of hibernation, it was much easier to hold more advanced cells at this stage than at other cellular checkpoints. (We could have used another, G1, but G1 gave us a window of opportunity lasting between half an hour and an hour. Use of this checkpoint meant having several teams working in a carefully choreographed way to coordinate all the elements of nuclear transfer.) We were ready to go.

FUSION

By now we have the nuclear donor and the enucleated egg, the CD player and CD, if you like. To pop the CD into the player and reconstruct the embryo for cloning, the DNA from one of these starved cells has to be introduced into our enucleated egg. One detail of all this that never fails to catch the eye of journalists is that, to kick-start development, we used a shock of electricity. Given how this happened in the winter months, there was the inevitable echo of the moment of Frankenstein's creation. ("It was on a dreary night of November that I beheld the accomplishment of my toils. With an anxiety that almost amounted to agony, I collected the instruments of life around me, that I might infuse a spark of being into the lifeless thing that lay at my feet.")

First, Bill would inspect the mammary cells and select a diminutive specimen, which was more likely to have been starved and thus near the start of the cell cycle. He checked that the cell had a normal surface, without a surrounding halo, another subjective sign of good health. Then he took the small pipette that he used for enucleation to pick up and pass the donor cell through the zona pellucida, usually through the same opening used for enucleation. The donor cell, like most cells in the body, is much smaller than the egg and fits within the pipette. He deposited the donor cell next to the enucleated egg inside the zona, which was enclosed in its own inner membrane (like the contents of a hen's egg under the shell). Because he had ten to twenty eggs in each chamber of the micromanipulator, Bill repeated

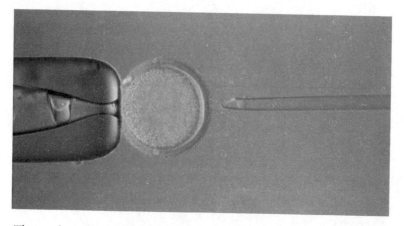

The much smaller nuclear donor cell is picked up by the pipette ready for placement inside the zona next to the enucleated egg. (Photograph courtesy of the Roslin Institute.)

this step ten to twenty times before taking them, en masse, to a second microscope to fuse their contents together, so that the egg could now "play" the new CD.

Fusion was then carried out by Karen Mycock. It took place in a special glass dish, an inch or more across and a quarter of an inch deep. This amazing feat of cellular intermingling was achieved on the bottom of the dish, where two wires were glued one-fifth of a millimeter apart. We had bathed the wires in a special salt-and-sugar solution, like the reconstructed eggs, so that they made a good electrical contact. First, Karen would prime the cells with a burst of alternating current, creating a different charge on the two cells, which drew them closer together until they kissed. After the AC came the DC. She passed pulses of electric current through the two cells to begin the process in which they would merge to create one. This is more difficult than it sounds, because other teams had shown, through trial and error, that to succeed the electric field must be perpendicular to the membranes where they touch. With the correct orientation the DC electric field drills holes in the cell membranes, enabling the contents of the cells to mingle.

This fooled the reconstructed egg into thinking it had been

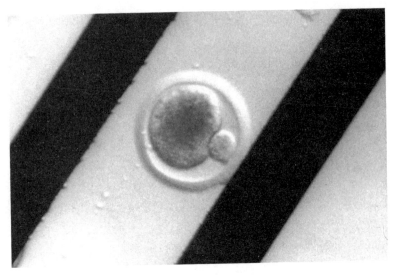

Electric pulses are used to fuse the two cells together. This demonstration with a mouse egg shows how the two cells are arranged so that, at the point where they touch, the cell membranes are perpendicular to the electric field that flows between the two wires. (Photograph courtesy of the Roslin Institute.)

invaded by a sperm, a process called activation that is linked to the release of calcium. In effect, Dolly—and you and I too—were conceived in a flash of calcium. After fertilization one can detect pulses of high calcium concentrations every few minutes that continue for several hours. In this way calcium triggers a chemical chain reaction throughout the fertilized egg that allows an embryo to form, and also to deny access to any other sperm by changing the properties of the zona, making it tougher, so that it can—in the body—resist the advances of any more sperm.

The calcium explosion throughout the egg's spongy interior also helps the successful sperm overcome a logistical problem. The egg is thousands of times bigger than a sperm, but the whole egg must undergo activation at fertilization. In the absence of the sperm, there are various ways to do this, for instance with a calcium bath or with an electrical current. To create Dolly, Karen used an electric shock to cause release of that all-important calcium locked up in cellular stores to trigger cell division.

After activation Karen put the reconstructed egg into the equivalent of intensive care: a culture medium where recovery could take place. When the contents of the enucleated egg and donor cell repaired the holes, the two cells merged into one. As Karen watched, she could see fusion gradually take place. Even after half an hour the smaller donor cell could still be seen nestling against the egg. After an hour or so the donor cell—and its DNA—had been swallowed up by the egg and its outline was not visible any more, rather as two beads of mercury become one. Dolly was one of the 70 percent of cases where the cells fused successfully. (In the wake of Dolly there is now an alternative technique for introducing the donor nucleus: the donor cell is sucked against the tip of a pipette that is smaller in diameter until the cell bursts. The contents are drawn into the pipette and then injected directly into the egg.) Karen and Bill would joke that it was they who were really Dolly's mum and dad. "I was the dad, and he was the mum," she has recalled.

Within a few days of reconstruction we expected the single cell of an embryo to divide several times to produce first two and then four, eight, and sixteen cells. Dolly was one of the 50 percent of embryos that seemed to be developing normally when all went well (sometimes not one embryo would develop). At the sixteen-cell stage, which develops after four days, the cells are independent but still held together by the zona to form a ball. The membranes of those cells that are on the outside of the ball stick to one another and begin to control the access of molecules to the cells in the middle. At this stage an embryo is known as a morula, and we begin to see the start of what really sets mammals apart from bacteria—the ability to divide the business of staying alive among a fully coordinated and cooperating community of cells.

Fluid begins to accumulate inside the morula to create a single fluid-filled cavity, at which time the embryo is a hollow ball of cells known as a blastocyst. There are between 50 and 200 cells in the

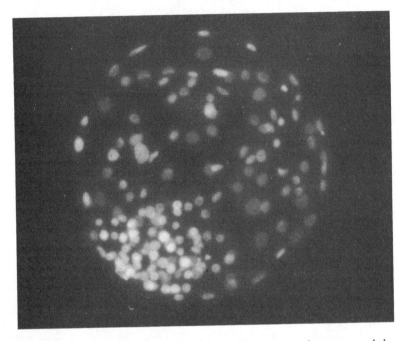

Sometimes we assess development of the eggs by staining them to reveal the nuclei, seen here as the small bright bodies. In a really good-quality blastocyst there will be two hundred cells. (Photograph courtesy of the Roslin Institute.)

embryo at this stage, depending upon the quality, and sometimes in preliminary experiments we stain the embryos to count the cells and make the first comparisons between different treatments. You can see the nuclei stained blue in these preparations. This is the stage at which an embryo can implant into the wall of the womb in some species, such as human and mouse. However, the ability to grow reconstructed embryos to this stage in the lab is by no means a guarantee that an offspring will result. In their experiments on mice, Teru Wakayama and his colleagues at the University of Hawaii found that varying the time between nuclear transfer and activation made little difference in the percentage of blastocysts that they grew, but only in those where there had been a delay did they manage to get cloned mouse pups. As ever, the details of how to succeed vary from species to species.

PLANTING EMBRYOS

WE WERE STILL groping for those important details when we created Dolly. We faced a key difficulty in growing reconstructed embryos up to the blastocyst stage, when they were mature enough to implant in the womb. There was no really effective method for keeping cloned sheep embryos alive this long in the laboratory. The medium that nurtured the embryos of the most studied species, the mouse, did not do so for sheep. The hard work to find the best recipe had not been done (teams in Australia and New Zealand led by Robin Tervit, Jeremy Thompson, and Simon Walker have now solved this problem). If we tried to implant embryos before they were blastocysts, they were destroyed because the process of enucleation and nuclear transfer had left them with a hole in their protective shell, the zona.

As ever, we turned to nature, exploiting another clever technique pioneered by that creative Dane—Steen Willadsen—which meant we needed a second set of surrogate mothers. To protect the reconstructed embryos from destruction, we encased them in a shell of agar. We tied off the oviduct (the equivalent of the fallopian tube in a woman) of another sheep and placed the encased embryos there, where they could develop for six days, to the blastocyst stage. This part of the operation would tend to fall on the Wednesday of a routine week.

The next Tuesday, when the embryo was about to outgrow its temporary shell, Bill had to recover the embryos and remove their coat of agar—one of the most difficult parts of the operation. Once he had stripped it of its protective shell, Bill could reimplant the blastocyst in a second ewe. This ewe also had to be receptive to the embryo: half a century ago, it was first shown that if you moved an embryo from one animal to another, the two animals had to be at the same stage of their reproductive cycle. There is a chemical dialogue between blastocyst and mother, and this must not be interrupted.

With the sheep under anesthetic, Bill's wife, Marjorie, made an incision just in front of the udder to expose the ewe's ovaries. To make sure the ewe was at the right point in her cycle, she would confirm whether the ewe had ovulated from both ovaries or only one. We tried to match the number of days that the cloned embryo had developed in the laboratory with the number of days that passed between the animal's being in heat and the surgery, then knocked off a day or two because reconstructed embryos did not manage to grow as vigorously as the real thing (still more evidence that cloning has deleterious effects). The jargon phrase is that egg and womb must be "synchronous." Work in the 1960s by Bob Moor and Tim Rowson in Cambridge showed that a blastocyst has the greatest chance of developing to term if it is transferred into a recipient ewe that has been in heat six or seven days previously, as if it had carried that blastocyst for that time. Then they are synchronous.

The development of embryos tends to be more successful in the part of the uterus that is next to the ovary from which an ovulation has taken place, perhaps because locally acting hormones help pave the way for pregnancy. After three or four days of being borne on the tiny waving fronds (cilia) that propel it down the reproductive tract, the embryo would normally reach the uterine horn. Because the embryo was less than six days old, the spot where it had the best chance of survival was at the tip of the horn. First we passed a needle through the wall of the uterus. Through the same opening we inserted a glass pipette carrying the embryo. The embryo was injected into the sweet spot within a small volume of medium. If all went well, at the age of about fifteen days, the transferred embryo would begin to attach to the wall of the ewe's womb, implanting by the fifth week. All we could do now was wait and watch the ewes.

WAITING FOR DOLLY

MY COLLEAGUE John Bracken, a soft-spoken man who has witnessed the birth of a dozen or so clones, would monitor the pregnancies with Douglas McGavin. Both had spent a lifetime looking after animals, particularly sheep, and knew and understood every detail of their care. For the Dolly project limitations of the ultrasound kit of the day meant that they could scan the sheep only two months into the pregnancy. Instead, they had to use old-fashioned methods to find out initially whether all the surgery and toil in the laboratory had paid off. They could tempt the surrogate mothers with "teaser rams" sixteen or seventeen days after the last time the ewe had been in heat: if the embryos had failed to develop, the animal would return to heat.

After a couple of months, John and Douglas could place an ultrasound probe against the belly of the surrogate mothers, the Blackface ewes; it was the very same equipment used for monitoring pregnancy in women. By this time the placenta had grown and the uterus of the surrogate filled with fluid, which could later be seen as dark patches on the screen of the scan.

In the first tranche of the experiments, we used cells from a nine-day-old embryo, the same type as those used to produce Megan and

The development of a fetus can be monitored by use of ultrasound scans, like those used with women to follow the development of their child. (Photograph courtesy of the Roslin Institute.)

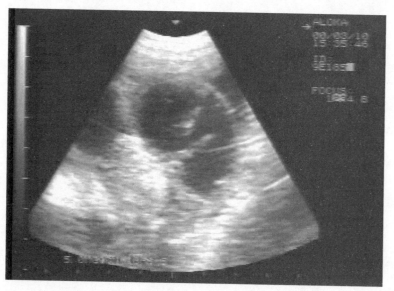

Morag, differentiated skin cells from a twenty-six-day-old fetus, and, of course, those frozen udder cells. The first ultrasound scans, at around eight weeks, showed that only one-third of the ewes had fetuses. Much to our surprise and delight, however, one ewe was pregnant with a clone from an adult cell. Ultrasound revealed an apparently normal fetus. Karen Mycock recorded the scan on a video with maternal pride and felt a thrill of anticipation that a lamb could result from this supposedly impossible experiment.

That day, 20 March 1996, was when we first began to think seriously about clones from adult cells. Keith and I became nervous, particularly as the pregnancies continued to fail. Except, that is, the one that should not have been: Dolly. We shared a mixture of excitement and concern because we knew that this would be a very important result, if the lamb survived. But our previous experience had shown that many of the fetuses die, sometimes late in pregnancy or even at birth. As we fretted over Dolly the fetus, the Princeton professor Lee Silver was writing a popular book on reproductive technology in which he was patiently explaining why cloning from adults was a biological impossibility.

That May, at around 110 days into pregnancy, four fetuses were found to have perished, all clones of embryos. Of the 29 embryos that we had created from 277 udder cells, the one fetus continued to thrive in her Scottish Blackface surrogate mother. We constantly monitored her well-being. Every time John Bracken used ultrasound, it always took a few seconds to get the big picture. You could usually see the head, legs, and ribs at first. There would also be some movement. Most important of all came the moment when we could see a heart beating. Then we all felt a great sense of relief and satisfaction.

To ensure that our nails were bitten to the quick, Dolly's gestation, like that of other clones, took longer than normal. In most British breeds of sheep the average duration of a pregnancy is approximately 147 days. Experience now tells us that pregnancy sometimes drags on up to 155 days for clones. They usually die if the pregnancy goes beyond day 153. The labor of a ewe carrying a clone is also unusually slow and sluggish, as we now know also to be the case in other species. Just before the onset of labor, a ewe having a normal pregnancy will often leave the rest of the flock and make a "nest" by pulling grass or straw together to form a hollow in a quiet corner of a field, an ancient reflex to create a defense against predators. For reasons we don't understand, nesting was not as likely to be triggered by a cloned embryo. The wait continued.

DOLLY'S BIRTH: MORE BLEAT THAN BANG

WE DID EVERYTHING we could to keep the ewes and the clones safe: we placed the animals under twenty-four-hour observation. We wanted to leave nature to take its course, but, if the pregnancy went beyond 153 days, we would induce birth with an injection of the same hormones that the fetus normally releases to signal it is ready for the outside world. The lambing pen at Roslin has seen great excitement and satisfaction but, more often than not in the case of clones, has also witnessed death and deformity. But that was not the case on 5 July 1996. A healthy lamb was born that day, one that would become the talking point of paupers and presidents.

Dolly entered the world head and forelegs first in a shed on the Roslin farm late in the afternoon. Her arrival was a muted affair. She weighed 14.5 pounds, surprisingly large for a normal lamb but not for a clone. Given what we now know—that the results of nuclear transfer are often oversize—that weight is, in retrospect, a pleasant surprise. In attendance were a local veterinarian and a few staff members from the Roslin. John Bracken was in charge. Even though I am often called the "father" of Dolly the sheep, I was an old-fashioned father and was not present at Dolly's birth. Her expected arrival was the cause of a great deal of excitement among the team, and I gave the instruction that only those who had to be present should be there. I, of course, obeyed my own guideline and was digging in the vegetable garden that I had on institute grounds. The last thing that we needed to do was impose stress upon Dolly's mother.

Bill Ritchie and Karen Mycock were thrilled with the idea that their cell manipulations had paid off. Bill was one of the handful of people present at the birth. "She was one of the few animals I saw being born," he said, recalling how he did "a bit of jumping up and down" when Dolly appeared. But, having made his first clone around 1993, and given that Megan and Morag were probably as significant, Bill admits that it came as an anticlimax.

Karen was at a friend's Highland wedding in Plockton that momentous Friday. Angela Scott had promised to keep her informed, since they knew the birth was imminent. When Karen returned to her room in the Craigendarroch Hotel to change her hat, she found a fax from Angela that read, "She's been born and she has a white face and furry legs."

"Heaven only knows what the receptionist thought as she slid the fax under our room door," she later recalled. Angela's message meant that the lamb was different from her Blackface surrogate mother. They could rule out any thoughts of a mix-up. They had a clone of an adult cell. Although Dolly had no name at that point, Karen knew that the lamb would make headlines. "I tripped the light fantastic. I thought hurrah and was absolutely ecstatic."

At the ceilidh in the village hall that night, Karen bought drinks for everyone. "I'm a dad," she shouted. "I think the locals will remember the 'mad scientist' who claimed to be a sheep's father and then sunk more than one tequila slammer and proceeded to dance the night away." Months later, Karen sent the bride and groom a copy of the Dolly paper so that they could understand what the fuss was about.

Keith, who had been so heavily involved in her creation, was not present for Dolly's birth either. He was with his partner, Ange, on a camping holiday in Devon that had been booked months earlier. When he rang in, as he did every day, and got the news, he cracked open a bottle of wine with her in a field to celebrate.

I did not hand out cigars or go for a celebratory drink in the local pub. No one took photographs. Nor could my wife remember my coming home and doing cartwheels of joy. Nor was Roslin painted red. Dolly's creation had been so long in coming, so protracted, so difficult, and such hard work that we were too fatigued to cry "Eureka!" I had bought a bottle of champagne for the event. But it seemed wrong to uncork it without Keith. In the end Alan Colman of PPL treated the entire team to a night out later in the summer when everyone was around.

There were other reasons not to shout about Dolly from the rooftops. Because we intended to submit the work for publication in *Nature*, we had to keep news of her birth secret for many months. The journal has a policy of rejecting papers if they are publicized before publication. The manuscript was submitted on 22 November and, after minor amendments, accepted in January, ready for publication toward the end of February 1997.

BROUHAHA

WE COULD SEE the media storm coming but were unprepared for its severity. Realizing that there would be a great deal of media interest, a public relations company, De Facto in Basingstoke, prepared a press release and other material. I practiced press interviews with Ron James, the biochemist turned entrepreneur who ran PPL, an immaculately tailored man with piercing blue eyes, white hair, and a white beard. We tried everything, from the aggressive jab of the microphone up the nose to the traps set in lengthy conversations with writers in search of a new angle ("So you really cloned Dolly because you wanted to clone yourself, then?").

As part of its normal prepublication process, *Nature* told reporters on Friday, 21 February 1997, that the paper was to be published on Thursday, 27 February 1997, triggering a trickle of advance calls. This practice of providing embargoed information allows journalists to investigate the subject, carry out in-depth interviews and be well informed before publication. By this time, the *Observer* newspaper—published on a Sunday—was already onto the story and ran it on the twenty-third, helped in part by a television producer whom we had told all about Dolly because his film was to be broadcast after her birth. The front-page story by Robin McKie announced, "It is the prospect of cloning people, creating armies of dictators, that will attract most attention."

I have no doubt that his hunger for a scoop made it more difficult for us to announce the result. We were only partway through a careful process of informing those people who might be asked to comment about what we had achieved: other biologists, ethicists, clerics—all those pundits whom the media turn to for guidance and comment. Doubtless they, like us, were being inundated by telephone callers requesting views on something they had not heard of until that Sunday morning. The premature publicity also caused difficulties within the Roslin Institute. We had planned to inform the staff members before the end of the embargo, but they heard from TV, radio, and newspaper about what had been going on in their own institute. One of our jobs early the following Monday morning was to address the staff to reassure them that we had no intention of cloning people, but had many other important uses in mind. We also apologized for the news leak and told them of what had been planned.

We found, despite all our hard work, that nothing could have prepared us for the thousands of telephone calls and the scores of interviews that followed. There were pronouncements from the great and the good. There was a great deal of angst. Reassuringly, there was little outright hostility and opprobrium. As soon as it was clear that the story was breaking early, Harry Griffin, the Roslin's bearded, affable, and much respected assistant director, and I went into the institute to take calls from the media. Ron James, who was in the south of England for the weekend, went to the Basingstoke offices of De Facto.

We could see the journalists of the planet wake up as Earth turned on its axis. First the reporters telephoned from Europe, then the East Coast of America, then the West Coast. Before the media hordes in Australia and New Zealand got up, we went home, wrung out and exhausted. Hundreds of e-mails began to arrive. The next day the Roslin was besieged by British and American TV crews. Satellite vans in the parking lot beamed live interviews to morning news programs on the other side of the Atlantic. Invitations poured in for talks at conferences, lecture tours, books, and films and even for Dolly to make a personal appearance on an American talk show.

The press conference, which we had been planning for Wednesday, was held at the Roslin on Tuesday morning, attracting a crowd even though the story was already "old," having broken days earlier. The story had legs, as reporters say. Four legs, in this case. With four film crews and twenty photographers, Dolly coped well, as if she had been doing media work all her short life. I was photographed with her many times over the years. Some city dwellers seemed not to understand that you cannot tell a sheep to stand in a specific place where their spotlights are focused, and so we began to use food to encourage her to come to a particular position. She was a sociable animal, even by the standards of sheep.

We were already beginning to tire of the incessant questions

Dolly at maturity seen in the paddock outside her barn, where she was often photographed, this time with Ian Wilmut. (Photograph courtesy of the Roslin Institute.)

about cloning people. They were repeated a thousand different ways, and many times even at the same press conference, but our answers were the same: We thought it was unsafe. We did not approve. We weren't interested. We became equally weary of cloning jokes: there was much play on William Blake's line, "Little lamb, who made thee?" Why, a significant udder, of course.

The next day photographers from *Time, Newsweek,* and *Scientific American* arrived. In a journalistic display of shock and awe, a trio of reporters from the *New York Times* flew in from Moscow, Germany, and New York. I was surprised to find that, at the age of fifty-two, I had become a celebrity. The *New York Times* described me in a section called "Man in the News": "His hobby is walking in the mountains of Scotland. He relaxes with 'a good single-malt Scottish whisky,' and enjoys the quiet of his village near Edinburgh that is 'so small you wouldn't be able to find it in an atlas.'"

All the time more TV crews were turning up, usually unannounced. A few weeks later Pepsi Cola offered my wife and me an all-expenses-paid trip in a private jet to a party with other luminaries in the Hotel de Paris, Monte Carlo. We accepted and were joined by Harry Griffin and his wife. I was also asked to appear in an advertisement for a European mineral water. I declined. A photocopier company came out with a cheeky advertisement that declared, "Big deal. We've been making perfect copies for years."

Our secretaries, Jaki Young for me, and Frances Fame for Harry, bore the brunt of the telephonic avalanche in the days after Dolly's birth. Still besieged by invitations and queries and radio stations across the planet, we were forced to draft in additional spokespeople. The Vatican and religious groups voiced their fears about cloning humans. President Bill Clinton asked his bioethics commission, led by Harold Shapiro, the president of Princeton University, to report within ninety days on the implications of our research. Dolly, the President said, "raises serious ethical questions, particularly with respect to the possible use of this technology to clone human embryos," and he called for a worldwide moratorium.

His statement carried strong echoes of the debate that greeted the birth of genetic engineering when, in 1975 at Asilomar, California, scientists in effect called a five-year moratorium on the new technology until biological understanding and regulation caught up. They were concerned that new techniques of gene splicing and cross-species genetic recombination might convert harmless microbes into virulent pathogens, among other things. Paul Berg, the Nobel laureate who proposed the moratorium, summed it up as follows: "When you honestly don't know the implications of your research, what are you supposed to do? Just keep on doing it? No, you have to stop and ask yourself what the hazards really are."

Even today I am still trying to work out the hazards of cloning. But on one point I have remained firm. Within days of Dolly becoming global news, I came to dread the pleas from bereaved families, asking whether we could clone their lost loved ones. I have two daughters and a son of my own and know that every parent's nightmare is to lose a child, and I know the lengths I would go to if there was any way to have them back. But I had and have no power to help. Cloning does not solve the problem of bereavement. Even if you did clone a twin of the lost child, it would be a different person, and that would be cruel to child and parents. For that reason, and for many others, I welcomed any move to ban the cloning of babies.

THE DOLLY DOUBTERS

THE COROLLARY of the surprise that greeted the announcement of Dolly's birth was the skepticism and the doubts. These were expressed by some eminent figures. I can't help feeling that they were fueled in part by the realization that Dolly was cloned by an obscure Scottish laboratory, rather than an American powerhouse of animal genetics. Then again, I would probably feel the same if a major advance in my field came from beyond my peer group.

The donor ewe was pregnant at the time her mammary tissue was taken, and that provoked a question from Norton Zinder, a distinguished microbiologist at Rockefeller University in New York City.

With Vittorio Sgaramella of the University of Calabria, in Italy, he asked whether it was possible that Dolly could have been the product of a fetal cell that had been circulating in the donor ewe's body and not an adult mammary cell at all. The *New York Times* tasted blood and scandal. Zinder had indeed made a good point, one we had not even thought about. However, PPL had tested circulating fetal cells to see whether a pregnancy was going to produce a truly transgenic animal. It had failed to find any, suggesting that Zinder's scenario was unlikely. And the Finn-Dorset cells from the Hannah were the only preparations from that breed in all the laboratories involved in creating Dolly. Nonetheless, never say never.

We were lucky: someone from the Hannah Dairy Research Institute rang up Alan Colman to say that they still had some of the mammary tissue used to make Dolly. My colleagues busied themselves with DNA fingerprinting tests to confirm that Dolly was the sheep we said she was. Our critics, in turn, demanded to know why these tests had not been required before publication in a peer-reviewed scientific journal of the stature of *Nature*. Scientists in my field of embryology were less rigorous than the molecular biologists who dominated the doubters, sniffed an article in the *New York Times*.

We upheld Dolly's provenance. Esther Signer of the University of Leicester provided independent DNA profiles confirming that Dolly was a clone of the six-year-old donor ewe. Signer, who worked with the inventor of DNA fingerprinting, Sir Alec Jeffreys, said that the results proved "beyond reasonable doubt that Dolly is indeed derived from a cell of the mammary tissue taken from the adult donor ewe."

There was another problem. At first Dolly was a "clone alone." An anecdote. An accident. The *New York Times* was troubled that no one had repeated the feat by the first anniversary of the announcement. In fact, this was unrealistic if you take into account the time taken to raise funds, buy the sheep, and then carry out the experiment. By the first anniversary there simply had not been enough time to reproduce the feat. Even so, in March 1998 Jean-Paul Renard and his team near Paris made it clear they had cloned a calf from differentiated cells, and that summer Ryuzo Yanagimachi of the University of Hawaii, and his postdoctoral colleague Teruhiko Wakayama, reported having cloned over fifty mice, including clones of clones, by nuclear transfer.

Since then research groups around the world have reported the cloning of species such as cattle, sheep, mice, goats, horses, pigs, mules, dogs, cats, white-tailed deer, carp, and fruit flies. Lacking, as we do, fundamental understanding of how to clone, there was a lot of trial and error to customize the Dolly method to suit each species. Successful teams found the feat difficult in some species, notably the rhesus monkey. But probably any mammal can be cloned, given enough research.

DOLLY'S LITTLE SECRET

DOLLY, LIKE so many other "clones," was not a true clone. She, like most other female clones, was made with an egg obtained from a female different from the one that donated the DNA for nuclear transfer. To have been a true clone, she would have had to have been made by means of an egg from the same long-dead Finn-Dorset sheep. Eggs contribute more to the development of the offspring than just the genetic information locked up in their chromosomes. Other genetic information is carried in the powerhouses of the cells, the mitochondria we encountered earlier, which lie in the cytoplasm of the egg.

Some 3 percent of the DNA in a cell can be found in the mitochondria. While the amount of information in the mitochondria is small in comparison with that held in the chromosomes, it is crucial for normal development. Mutations in mitochondrial DNA accumulate with age more quickly than in nuclear DNA—for instance, as a result of the action of highly reactive free radical chemicals that are naturally produced in the body. If cloning affected mitochondrial DNA, it could be significant, in the wake of a study by Nils-Goran Larsson and colleagues at the Karolinska Institute, Stockholm. He bred mice with a defective version of DNA polymerase, an enzyme responsible for maintaining mitochondrial DNA, so they suffered more mutations. These "mitochondrial mutator" mice live only ten to twelve months, compared with up to three years for normal mice, and they aged prematurely.

Abnormalities in function of mitochondria can also lead to disease. Richard Lifton of Yale University School of Medicine found a genetic defect in mitochondria linked with hypertension, high cholesterol, high triglycerides, low magnesium, diabetes, insulin resistance, and obesity (the defect affected transfer RNAs, critical carrier molecules that ferry amino acids around during the construction of proteins, contributing to a broad range of cellular malfunctions). Given their role in powering the body, the first manifestations of having aged mitochondria would be in organs that have a high demand for energy (and are thus more dependent on mitochondria), such as the heart and brain.

Although we have yet to discover all the details of how the egg's mitochondria affect the clone, there is plenty of indirect evidence that they are important. Take a number of eggs from the same ewe and, after adding sperm and reimplanting, you will find that some produce healthy lambs, some do not develop at all, and others do develop for a little while but ultimately the embryo dies. Sometimes these differences in the viability of eggs will be down to differences between their mitochondria. But they are not the whole story. Aside from the influence of the mitochondria, events in the earliest stages of embryo development are also influenced by proteins and other molecules in the egg that were made before the egg even left the ovary. Very few of these active compounds have been identified, nor is their function understood.

BEYOND DOLLY: KAMITAKAFUKU

KAMITAKAFUKU IS THE Japanese for "God-given high happiness," as well as the name of a Black bull who was highly prized for his marbled, fatty "Wagyu" meat. By the time he was seventeen, his semen had been used to inseminate more than 350,000 cows, and this prodigious father had become a star in Japan, where bulls of his caliber are worth a fortune. In a conventional breeding program, it may take seven years to select a genetically elite stud bull like Kamitakafuku from hundreds of his peers, because of the lengthy process of testing for desirable traits. Rather than purchase the semen of genetically elite bulls to inseminate hundreds of cows in the hope that calves will go on to develop these traits—despite the dilution of these elite genes with those of another animal—the beef industry hoped instead that cloning would assure valuable copies of elite bulls without all the testing and selecting of progeny. By the same token, cloning could also be used to speed up the development of breeds suited for harsher conditions in developing countries.

Kamitakafuku died in 2001. But his genes lived on in more than the hundreds of thousands of offspring that he had sired. Kamitakafuku was the subject of an early mammalian cloning experiment that began in 1997, the year after Dolly was born. Xiangzhong (Jerry) Yang, of the University of Connecticut at Storrs, and Chikara Kubota, of Japan's Kagoshima Prefectural Cattle Breeding Institute, removed ear cells from the bull for nuclear transfer and in 1998 reported the births of six offspring. Four calves survived, including the first male clones of a farm animal. This initial set of cloned offspring—Saburo, Taro, Jiro, and Shiro—developed normally into adults. Saburo and Shiro fathered offspring through artificial insemination, as well as the old-fashioned way. Then, using ear skin cells from these clones, the team once again performed nuclear transfer. Nineteen embryos were transferred to nineteen recipients, which led to two live births of second-generation cloned bulls on 23 January and 6 March 2000. Although one of the two calves died of anemia and infection shortly after birth, the second lived on in apparent good health. The survivor was given the name Sho-zaburo and the English name Dr. Lederman, after Jerry Yang's doctor, who routinely joked he would like to be cloned.

The story took another twist in 2005 when Saburo and Shiro were slaughtered and their meat tested. Understandably, Jerry Yang was reluctant to do this, given all the effort his team had put into their birth in the first place. But, because of the abnormalities caused by cloning, some scientists had raised theoretical concerns about the safety of eating clones. Yang's pilot study was the first to examine specific proteins and nutrients in the meat from cloned animals and marked the start of efforts to fill an important gap in the research literature and achieve regulatory approval of clone-derived food. With Cindy Tian, Chikara Kubota, and colleagues, Yang had also cloned a particularly productive Holstein dairy cow (called Aspen), then compared the meat and milk from the clones to that of animals of similar age, genetics, and breed created through natural reproduction.

"We conducted extensive comparisons of the composition of milk and meat from somatic [adult cell] cloned animals to those from naturally reproduced comparator animals," he said. "We found no significant differences." Although many more studies would be necessary to convince regulators, the data provided the first science-based information to address public concerns about the safety of meat and milk from somatic animal clones.

This kind of work is also encouraging when it supports the hope of scientists, like me, that breeding can be made more precise if cloning is combined with genetic modification. Inserted genes offered a way to enhance breeds, adapt them to different environments, or alter them so that they could grow "humanized" rejection-resistant organs for transplant. They could even be designed to manufacture precious drugs in their milk. One could clone from cultured cells that had been transformed genetically to make a drug as they were grown. Beyond Dolly an exciting landscape of new scientific opportunities beckoned.

FOUR

FARMYARD
CLONES

OUR FIRST ATTEMPT to blend the techniques of cloning and genetic modification rested in part on the solution to a bloody mystery that was published half a century earlier, when a paper appeared in the 1952 Christmas edition of the *British Medical Journal*. Working in Oxford University, Rosemary Biggs, Gwyn Macfarlane, and their colleagues had found a new form of hemophilia, the inherited disorder that causes recurrent, uncontrolled bleeding.

An important clue that there was more than one form of the clotting disorder emerged when it was found that the blood of some sufferers could clot the blood from a patient who had the more common form, hemophilia A. Later it was discovered that this variation of hemophilia was caused by the lack of a different clotting protein, one called factor IX. The Oxford team decided to name this new form of hemophilia Christmas disease after their very first patient, a five-year-old boy called Stephen Christmas, who had by then moved to Toronto in Canada. The journal rushed to get the paper into print at the appropriate time.

There was a precedent for naming disorders this way, first set by the British surgeon Sir Jonathan Hutchinson. Indeed, Stephen's father, an actor, had himself played with the Yuletide theme. He named his older son Robin and, when he remarried, named the daughter of that union Holly. But, given that seasonal papers in the Christmas edition of the *BMJ* were traditionally flip and light-hearted, some doctors were unhappy. They had been expecting a meditation on overeating, drinking to excess, or some festive frivolity. Some said that to call such a serious condition after a Christian festival was tasteless. Others dismissed the paper on Christmas disease as a student prank. One professor wrote to the journal to complain: "Some parents live to regret the names they have thoughtlessly bestowed upon their innocent children. Would it not be possible . . . in your capacity as registrar of its birth, to substitute some less ridiculous name?"

The Oxford team stoutly defended the name but did assure the critics that if they found the protein used in the body to make factor IX, they would resist the urge to call it Christmas Eve factor. The only alternative name that the Oxford team offered was the dry, functional "hereditary hypocoprothrombinemia." Other candidates were "parahemophilia" and "pseudohemophilia." Unsurprisingly, "Christmas disease" stuck, though today we tend to use an alternative, "hemophilia B" (suggested by a distinguished scientist during all the arguments). Unfortunately for Stephen Christmas, however, there was only one way to get hold of the clotting agent that he lacked: by means of human blood.

Half a century later we found a different way to make factor IX, in the milk of a sheep called Polly. Her birth marked the first demonstration that cloning would also make it easier to carry out the genetic alteration of animals than ever before. In many ways her birth thrilled me as much as that of Dolly. Once again our work was controversial since there are many who believe that Polly provided another example of scientists' "meddling with nature." Others found the idea of putting human genes into animals disconcerting.

But I felt, as I am sure did all of my colleagues in the Roslin and PPL, where Keith Campbell was working by the time the details of the feat were published, that our attempt to turn Polly into a woolly drug factory was a natural extension of many efforts by people to make useful products from biological processes—biotechnology—that dates back thousands of years, to the fermenting of wine during the Neolithic era, in the Middle East and in China. The aims of the early farmers were little different, in spirit, from what we wanted to do. I would also like to think that Stephen Christmas would have approved of our objectives.

When we unveiled Polly along with four other cloned lambs, we felt we had passed an important milestone in the long effort to adapt animals for life on the farm, creating new opportunities in the process. Indeed, we hoped that Polly's arrival would mark one of the biggest advances in breeding since the origins of animal domestication many millennia ago, because the techniques used to make her rendered it possible to alter animal DNA in a more precise way than ever before. In parallel, new ways to alter the way genes are used in the body have also emerged.

Since the birth of Polly we have seen cattle able to make human blood proteins, pigs with organs customized for transplant into people, chickens that lay eggs laced with anticancer antibodies, and goats that make spiderweb silk. Despite all this promise the full impact of this technology has yet to be felt. The techniques of genetic modification that I describe below could do even more for human medicine, a theme I want to return to at the end of the book.

WHY POLLY?

OUR NEIGHBORING biotech company, PPL, was keen to build on Roslin's success with cloning to try genetic modification, along with rivals elsewhere, notably in North America. By the time we started working on Polly, the use of genetically altered organisms—bacteria and yeast—to make proteins had become routine. These were no ordinary proteins but ones that could be used as medicine.

Called recombinant proteins, they are used extensively in the food, paper-processing, and pharmaceutical industries. Among the first to be produced by yeast were human growth hormone, used to combat dwarfism; human insulin, for diabetics; and the interferons and lymphokines, such as interleukin-2, which boost the body's immune system.

However, farm animals like Polly offer many advantages over microbes when turned into living drug factories. Bacterial fermenters—the huge stainless steel vats in which they were grown—were much more expensive to run than feeding sheep hay in a field. There was, moreover, a deeper problem in using microbes: these simpler organisms can make only simple proteins. More complex proteins found in people carry sugar residue molecules that are necessary for them to work properly in the body. Bacteria and yeast are not able to add these sugar groups, and before pharm animals appeared, such complex "glycoproteins" had to be isolated from blood or human tissues, which was an expensive process and risked isolating human diseases too.

We were not alone in becoming excited by the possibilities of transgenic animals since the first had been created by microinjection in the 1980s, culminating in 1990 with the unveiling of Tracy. There were many other researchers racing to become successful "pharmers," such as James Robl at the University of Massachusetts, Amherst, Randy Prather at the University of Missouri, Columbia, and Neal First at the University of Wisconsin, Madison. Companies other than our neighbors and collaborators, PPL Therapeutics, included Advanced Cell Technology, GenPharm, Genzyme Transgenics, and Alexion Pharmaceuticals in America and Pharming Holding in the Netherlands. If scientists could carry out precise "genetic surgery" on an animal, the commercial possibilities would be huge.

From the milk of genetically altered sheep the Roslin and PPL team hoped to harvest the factor IX that Stephen Christmas lacked.

Others were racing to pharm antithrombin, a protein that helps to stop blood clotting; human albumin, for use in the treatment of burns; proteins that could boost the numbers of red blood cells to treat anemia; or proteins to treat osteoporosis, arthritis, malaria, and HIV. The hope was that we could make these proteins at a fraction of the conventional cost. A traditional protein-drug factory would cost $200 million to $400 million and take between three and five years to build. By comparison a new strain of sheep would cost $100 million to develop and take eighteen months to start lactating so we could extract drugs from their precious milk. In its lifetime one animal could make $200 million to $300 million worth of pharmaceuticals, depending on the protein.

Apart from turning sheep into "bioreactors," as some like to call them, the introduction or deletion of genes also offered the means to make animals that grow faster for meat production or are resistant to diseases. Bovine Spongiform Encephalopathy (BSE) and scrapie were obvious targets, for instance. Genetic changes could benefit the environment; one example was the "enviro pig," a swine that had a phytase gene placed in its salivary glands so that it made more use of phosphorus in feedstuffs and less phosphate ended up polluting the environment. Some wanted to genetically alter animals so that they could produce human milk, a rather far-fetched idea.

Several labs, including our neighbors, PPL Therapeutics, hoped that transgenic animals would eventually provide suitable organs to overcome the desperate, and chronic, shortage of suitable organs for transplant. At present thousands of people die each year before a suitable organ becomes available. The company wanted to modify pigs so that their hearts or kidneys could work in human patients. The technology could also be used to make animals designed to suffer humanlike diseases for testing treatments, an application that has seen a huge surge in the use of genetically altered mice. Other companies wanted to use transgenesis to enable goats to make spider silk protein in their milk, paving the way to a new generation of wonder fabrics. The possibilities were, and remain, endless.

Quite predictably, animal rights campaigners stepped up the

pressure to study the welfare of transgenic animals and examine the ethics of tailoring animals for human use. We were ourselves uncertain of the long-term effects of the procedures used to create Polly. She was a first, after all. However, concerns about animal welfare should, I feel, be kept in perspective. Scientists and companies must think carefully about whether their work is justified: no one wants to modify animals for trivial reasons.

There was also unease that we had somehow "humanized" sheep. Polly was "transgenic" and had a human gene in every cell of her body, after all. I should stress, though, that she was still very much a sheep: sheep are thought to carry around 25,000 genes in all, similar to the human complement of genes, and just one extra gene, even if it was "human," would make little difference to her "sheepness" defined by the other 24,999 genes. Indeed, all people are somewhat sheepish. If you compare the genetic code of human and sheep, the DNA sequences of sheep genes are about 85–95 percent identical to the corresponding human genes, with the sequences of some genes more highly conserved than others. Given that a human gene itself was not used, but a clone (a copy) of that gene, Polly was a long way from being a humanized sheep.

GENE TINKERING

BY THE TIME we started the Polly project, there were many ways to introduce the DNA of foreign genes into the cells of an animal. You could add calcium salts to make the DNA that codes for a gene precipitate out of a solution onto the embryo and hope that some would penetrate to the nucleus to add to the DNA already there. You could use electrical pulses to punch small holes in the membranes of the embryos and to drive DNA inside. You could package DNA in fat particles (liposomes), which are soluble in the lipid membrane around cells, so they can pass inside them. The most accurate method, also the most labor-intensive, involved the injection of a few hundred copies of the gene directly into the nucleus of a one-cell embryo, a zygote. The last was the method the Roslin and PPL used to make their sheep. Typically, of 10,000 embryos injected with foreign DNA, only 3 would make it to productive adulthood. Imagine the slog of doing something 10,000 times to achieve just 3 successes.

There were many reasons why this early kind of DNA modification was a lottery. Often the embryos that were injected stopped developing. Less than 20 percent developed to term. Sometimes injected DNA failed to become incorporated into the genetic code (genome) of the zygote until the developing embryo had formed up to eight cells. Then we would find DNA in some cells of the developing animal, but not in others, forming a mosaic of modified and unmodified cells, a chimera. To make matters worse, the novel DNA can be incorporated randomly in an animal's genome, and this can have unexpected effects, interfering with the action of other genes, or regions that turn genes on and off.

Even if the gene was incorporated and not causing damage, there was plenty more that could go wrong. The gene may be in the genome but not used properly. The way genes are used in the body depends on location. A gene that makes a cell hearty should be used only in the heart. Similarly, a gene that makes a nerve cell brainy should be used only in the brain. The same holds for milk-producing genes and mammary glands. There is a danger that a newly introduced gene can be turned on in the wrong tissue or organ. This is what damaged the "Beltsville pigs," when an American team's attempt to boost the growth of swine with a human growth hormone went awry—the hormone was made and used in the wrong tissues. The result was arthritis, lameness, mammary development in males, and a host of other problems. And a lot of bad publicity for this technology.

To turn a gene on in the right place—where you want it to act— you need a gene switch called a promoter. In an ideal world the promoter should be one that already controls genes in the animal you want to alter genetically. To create a pharm animal that could make the drug AAT (alpha-1-antitrypsin) in her milk, we thus attached the AAT gene to the promoter region of a gene that is responsible for a sheep milk protein, called beta lactoglobulin. The promoter ensured that it was used only in her mammary gland, even though every cell in Tracy's body had the AAT gene. That way her milk contained a protein that was being used to treat cystic fibrosis and emphysema, but without the expense and the risk presented by AAT from human blood plasma.

Tracy required a lot of effort to produce. This being the days before cloning, we injected a thousand embryos with this "construct" of gene and promoter. We implanted them in around a dozen sheep a day over one breeding season. This required two people to anesthetize the sheep, two to operate on the animal, one to remove the eggs, and one to do the gene injections. This was a huge enterprise. Fortunately, it worked. In 1990 Tracy the sheep was born. Tracy was outstanding, a pharming pioneer that could make 30 grams of the human protein for every single liter of milk that she produced. A good dairy sheep will make 900 liters during its five- to six-month lactation period. Some breeds have three lactations in two years. That meant she could make 45,000 grams per year, a vast amount when you bear in mind that around 200 grams per year were required to treat a single patient. In other words, Tracy could supply around 200 patients all by herself.

Until Dolly was born, Tracy was the most famous sheep in the world. She was then followed by other pharm animals such as Herman the bull, created by GenPharm's Dutch subsidiary. Herman (named after a company vice president), carried the human gene for lactoferrin and became the father of at least eight calves in 1994. Each one inherited the gene to make an iron-containing protein that is essential for infant growth, and the developers hoped that with them a new source of nutritious milk could become available. There was Grace the goat, which produced an anticancer protein. At the time of writing, however, the first product to treat human disease to reach the market looked likely to be GTC Biotherapeutics' (formerly Genzyme Transgenics) antithrombin III agent, ATryn, a blood protein with anticoagulant and anti-inflammatory properties, which is produced in the milk of transgenic goats.

Genetic modification can be of direct benefit to farm animals. We have seen the first transgenic cows that resist disease: mastitis, an infection of cows' milk glands that is often caused by the bacterium *Staphylococcus aureus*. To combat infection, a team from the U.S. Department of Agriculture introduced a gene from the related bacterium *S. simulans* into Jersey cows that allows them to produce lysostaphin, a protein that kills *S. aureus*. In time other strategies will emerge to provide animals with greater resistance to disease.

POLLY

POLLY WAS MADE in a different way from Tracy and other early pharm animals, one that promised to be much less hit and miss. My team at the Roslin—with Angelika Schnieke (the first author of the *Science* paper describing our method), Alex Kind, Karen Mycock, Angela Scott, and Alan Colman of PPL—started with tissue taken from a Poll Dorset sheep. Copies of the fibroblast cells in connective tissue taken from this animal were grown in the laboratory; then human genes were introduced. They were labeled with a marker. Typically, these markers protected the cells from drugs that would otherwise have killed them. In this way their survival showed that the human genes had been taken up and were functioning.

In the case of Polly, we introduced the human factor IX gene, along with DNA sequences that protect cells from the antibiotic neomycin. Thus, we reasoned, any cells that become resistant to neomycin after modification must also contain the human gene. This feat could be used to sort the cells for those where genetic modification had worked. This was the important step that spared us from having to transfer vast numbers of transformed embryos into ewes to find out whether the genetic alteration had been successful. Now we could take the nuclei of the selected cells—the part containing the genetic instructions for the Poll Dorset sheep and the human protein—and use them to "reprogram" sheep's eggs from which the DNA had been removed by the nuclear transfer method. By 1997 the process of cloning had become so routine that Karen can remember little of the effort to make Polly. "We had become a production facility and were churning out the numbers."

The resulting embryos were transplanted into Scottish Blackface ewes. Pregnancy was easier to achieve than in the case of Dolly. While only 1 of the 277 earlier attempts succeeded, we achieved a success rate around 1 in 50 (2.25 percent) in the various campaigns to make Polly and her peers. This was encouraging and suggested that the transplant of a human gene had not disrupted the development of the fetus. Whereas Tracy required around 50 ewes to create her, Polly required only around 20. And we did not waste any resources on creating nonmodified lambs, an expensive, pointless, and roundabout replacement for normal procreation on the farm (although this higher initial efficiency is, of course, complicated by the reality that fewer cloned embryos are viable and some cloned animals are unhealthy).

Blood samples were taken from the resulting lambs and used to confirm the presence of the added genes. Of the five lambs, one—Polly—had the human gene plus a marker gene. The newly introduced gene enabled her to make factor IX. PPL was obliged to tell its shareholders about this in advance of publication, so *Nature* huffily refused to carry the paper. It ended up in the equally prestigious journal *Science* toward the end of 1997. In the same issue the birth of Dolly was celebrated as the scientific breakthrough of the year. When Polly stepped onto the scene, earlier work on Tracy seemed to have paid off and PPL had been given permission to assess the use of the sheep-derived human protein AAT to help prevent lung damage in patients with cystic fibrosis.

THE ALTERNATIVE

BEFORE THE USE of genetic engineering, Stephen Christmas and other suffers of hemophilia relied on blood donors to help survive. Large volumes of blood were needed to provide the level of factor VIII or factor IX required to control a bleeding episode, so there was always a risk of heart failure. The ability to transfuse patients with plasma rather than with whole blood represented a step forward, though large volumes were still required. Then, in the mid-1960s, it was discovered that as frozen plasma was thawed a portion would remain in the form of sludge that was rich in factor VIII (but not factor IX). This led to the development of a product known as cryoprecipitate, but it did little for Stephen.

The biggest impact on Stephen's life was to come from the supply of his missing factor in a more convenient form. Sufferers of the more common hemophilia were the first to benefit when a concentrated form of factor VIII was made from large pools of plasma. Known originally as factor VIII concentrate, it was in the form of freeze-dried powder that could be reconstituted with sterile water and then injected. These concentrates revolutionized hemophilia care. Hemophiliacs were now independent of hospitals and could hope to lead normal lives. Soon a way to isolate factor IX from plasma was also developed. In this way Stephen Christmas could rely on a convenient concentrate made from the plasma collected from blood donors. He had a passion for photography and became a medical photographer at the Hospital for Sick Children in Toronto, although he worked mainly as a cab driver. In this job it was easier for him not to turn up for work if he suffered bleeding into his joints.

The first step toward severing hemophiliacs' dependence on human blood came when scientists found and characterized the human genes for factor VIII and factor IX, the instructions used by the body to make these proteins. The feat was pulled off for factor IX in 1982 by researchers including George Brownlee and colleagues at Oxford University and for factor VIII in 1984 by Ted Tuddenham's team at the Royal Free Hospital Hemophilia Center in London, working with Dick Lawn's team at Genentech in San Francisco, and by Jay Toole's at the Genetics Institute in Boston. These advances made it possible to use these genes in genetic engineering and, in turn, to create Polly, and thus alternative supplies of so-called recombinant proteins, as well as to pursue novel treatments based on gene transplants.

GENE SURGERY

EVEN THE PIONEERING WORK on Polly was far from being an ideal example of how to use genetic technology to alter the genetic code of an animal, however. We still crossed our fingers in the hope that the human DNA we had dropped into her cells actually worked. And if it worked, we held our breath that decent amounts of protein would be made, that it worked in the right tissues, and that plonking a new gene into the genome would not disrupt the workings of other genes. This happened in around 7 percent of our attempts. Even if only one of a pair of chromosomes was damaged this way, inter-breeding between two such sheep could produce lambs where both copies of the chromosome were damaged, along with the prospect of a healthy life.

All this uncertainty spurred efforts to refine the alteration of the genetic code of mammals, from sheep to people. We wanted an easy method of targeting genes in some way to the right place in an animal's genetic code, rather than simply hoping that a gene can be parachuted into a creature's genetic code without mishap. One kind of "genetic surgery" was achieved by my colleagues at PPL in two other cloned lambs, Cupid and Diana. Employing a technique that had been used for years on mice, but with little success on other mammals, Alan Colman, Alexander Kind, and the rest of the PPL team inserted a gene that coded for a human protein into sheep fibroblast cells—the flat, elongated cells found in connective tissue. They used a method called gene targeting, where they sandwiched the DNA sequence they want to insert with snippets of code that complement those on a target site. These "flanking" regions of DNA exploit a built-in feature of DNA's double helix: certain "letters" of DNA code like to stick to each other. The trick was to match the DNA "bookends" that surround the genetic material being inserted with the DNA on the target site. Because of the tendency for similar strands of DNA in the bookends to be attracted to those in the target genome, from time to time, as they entwined, there was an exchange of DNA sequences, a process called homologous recombination. In this way the researchers targeted a DNA sequence into a specific region on a chromosome of a sheep cell.

They chose a site for insertion that directed the production of one of the proteins that make up fibrous tissues such as cartilage, a collagen gene. They selected this gene because it is expressed in a great variety of different tissues and therefore was likely to be a site in which reliable functioning of the added gene would occur. Even with the help of gene targeting, the gene was expected to end up in the right place only once in 10,000 attempts. As before, an antibiotic resistance gene was added too: then the cells where the gene surgery had worked flawlessly were identified by their resistance to an antibiotic and selected for use in cloning.

Although this work was effective and took the project a step forward, we have rethought the practical value of nuclear transfer since the experience with Cupid and Diana. While fewer animals are used overall, the welfare problems linked with the present methods of nuclear transfer—to which I will return later—outweigh this advantage. We decided not to use nuclear transfer simply to add genes, reverting to the old-fashioned method of direct injection. However, cloning can still be justified to change genes, since we cannot do this any other way.

You could, in principle, eliminate the kind of genes responsible for the degenerative spongiform diseases, BSE and scrapie. At the Roslin we tried to do just this. By then the teams had grown with new recruits in both the embryology and the molecular biology labs. A team led by John Clark sought to make animals resistant to BSE in cattle or scrapie in sheep. These and other spongiform diseases result from mutations in a gene that directs the production of a specific protein within the brain, the "prion" protein. In sick animals the protein changes shape and triggers a molecular "domino effect," converting normal prion into the abnormal form that accumulates into sheets that disrupt the function of the nerve cells. Experiments in mice showed that if the native prion protein is not produced, there can be no spongiform disease.

Inspired by this finding, we set out to knock out the prion protein in sheep, which can suffer from scrapie. Our first objective was to prevent the gene from functioning and to show that the animals were still perfectly healthy. Then they would be exposed to abnormal prion to see whether they would succumb to scrapie. A single lamb was cloned—the first sheep with a specific gene deletion. It was not healthy, however. Although the animal was very strong and active and had a big appetite, it panted all the time, even when resting. We did a very thorough investigation and asked advice of veterinarians and a local hospital, but to no avail. After twelve days we decided that it was kinder to put the lamb down. Detailed histopathology, carried out by Susan Rhind at the local veterinary school, showed abnormalities in the blood vessels that supply the lungs. This was the first time that this abnormality had been seen, and it provided yet another salutary lesson that cloning a person would be very dangerous at present. Nevertheless, we had taken another significant step, and there were other important objectives. Unfortunately, the Geron Corporation withdrew its funding for the research project at the Roslin, and we became mere spectators of what was to follow.

HUMANIZED PIGS

THE OPPORTUNITIES for genetic modification have expanded as each new species has been cloned. On 5 March 2000 five piglets were born, the first swine created by nuclear transfer from adult cells. When PPL unveiled the litter, it said that nuclear transfer has proved to be more difficult in pigs than in other livestock, in part because pig reproductive biology is inherently more intractable, and in part because pigs need a minimum number of viable fetuses to maintain pregnancy (sheep, cows, mice, and humans need only one). The latter problem was overcome by tricking nature: the PPL team introduced parthenogenetic ("virgin birth") embryos with the clones. The parthenotes would not develop to term, competing all the time for limited resources, but would survive long enough for the pregnancy to succeed.

The first piglet, Millie, was named after the millennium; Christa, after Christiaan Barnard, the surgeon who performed the first human heart transplant, in 1967; Alexis and Carrel, after the transplant pioneer and Nobel Prize winner Alexis Carrel; and Dotcom to reflect the growing use of the Internet. The work was partly funded by a U.S. government grant awarded to produce a "knockout" pig—one that has a specific pig gene or genes inactivated to help prevent the human immune system from rejecting an implanted pig organ. With the piglets PPL had shown that cloning was possible and said at the time that it had already achieved the required targeted gene knockout in pig cells, using the same technology that led to the gene-altered lambs Cupid and Diana.

This put pigs at the heart of the effort to produce organs for "xenotransplantation," the transplanting of animal organs into people. Apes would, of course, have been a closer match, but there were troubling ethical concerns when it came to using a fellow primate, and the fact that they produce only single offspring would have made efforts to humanize them hard work. Also there would be a bigger risk that infections would spread from the animals to people—not just to the patients, but to the entire population. Meanwhile, the reality that pigs have large litters and that their organs are similar in size to human organs has led to their being the most likely suppliers of any future transplant organs.

Although the size of a pig heart is right for transplant, as is pig anatomy and blood pressure, it turns into a blackened mess in a couple of hours if human blood is pumped through it. This "hyperacute rejection" is initially triggered by antibodies in human blood and then executed by a cascade of enzymes that attacks the membrane of foreign cells. One pig gene in particular offered the biggest obstacle to pig-human transplants, triggering human recipients to reject the animal's organs. The gene, alpha-1,3-galactosyl transferase, is responsible for making an enzyme that adds a sugar (alpha gal) to the surface of pig cells that is recognized by the human immune system as foreign (humans and Old World monkeys lack the enzyme and thus this sugar coating). Without the sugar, human antibodies cannot attach to the pig cells, and therefore the acute rejection process does not begin. The race was on to produce a genetically modified pig that lacked the gene.

Two years later, PPL's subsidiary in Virginia announced the birth of five healthy cloned piglets that had one copy of the tissue rejection gene knocked out. Because the piglets were born on Christmas Day, they were given seasonal names: Noel, Angel, Star, Joy, and Mary. The company thought that a jolly name delivered more column inches and made scientists look more caring and sharing than calling their experimental subjects porcine one, porcine two, and so on. At the time PPL seemed to be the first to achieve the feat, but a litter of humanized swine had already been produced by a rival American group consisting of Randall Prather of the University of Missouri, Columbia and Immerge BioTherapeutics, a company created by the merger of the U.S. company BioTransplant and Imutran—the UK start-up company from Cambridge that led the race toward xenotransplantation before being bought up by the Swiss pharmaceutical giant Novartis. Their humanized swine had uttered their first squeals in September of 2001, and the team had gone to the trouble of submitting its work for peer review in the journal *Science*, which meant it had to keep quiet until its work had been peer-reviewed and published.

PPL had announced at the time of its festive litter that it expected the first application to be in the production of insulin-producing islet cells from knockout pigs for the treatment of diabetes, first in primates and soon thereafter in humans. Clinical trials could start in four years, it said, citing analysts' belief that the market could be worth over $5 billion for solid organs alone and $6 billion for cellular therapies for diabetes, Parkinson's, and Alzheimer's. The most headline-grabbing use of these pigs would have been for heart transplants, but there were—and still are—huge obstacles to overcome.

At the time of the announcement of the deletion of one copy of the gene, the rival teams had to do more work to knock out both copies of the alpha gal gene in each pig. It turned out that it was not necessary specifically to make the change. Spontaneous mutation in the remaining "wild type" copy of the gene occurs frequently enough to make it possible simply to select those cells. Cells in which the gene was present and functioning were killed selectively by introducing either a bacterial toxin that binds to the sugar or an antiserum against the galactose residue on the cell surface. Surviving cells were used as nuclear donors and yielded pigs that lacked either copy of the gene. The two groups published their findings, PPL in 2003 and the Missouri group in 2004.

Simple enough. However, the overall objective of a double-knockout pig in itself still seemed unlikely to guarantee success: the human body rejects transplanted tissue as a result of both short-term and long-term attack by the immune system. This can lead to the buildup of deposits, akin to those seen in heart disease. The details of this "cell-mediated immunity" are poorly understood.

Tests are needed to see exactly what effect the gene deletion has had on the pig tissues. The animals themselves are healthy enough, but what happens if blood from a human or another Old World primate is passed through the tissues? Initial results, recently published, suggest that the tissues are healthy for a prolonged period, much longer than the minutes of a normal pig, but much remains to be learned. Both companies claimed to have answers. Immerge said that it was looking at ways to reduce the immune response in patients to complement its GM efforts, a strategy that had a downside in potential side effects. PPL wanted to go farther down the GM route, adding three more genes, such as DAF (decay accelerating factor), as well as genes to prevent unwanted coagulation within the blood vessels of the transplanted organ, to address other stages of rejection. But no one knew how many genetic changes were enough to make a pig organ sufficiently human that it could still beat properly and pump blood in a human body without being attacked by its host. Some researchers estimated that as many as six hundred genes may be involved in tissue matching and warned that there could be a limit to how much pig tissue can be humanized—no one knew whether, if you altered enough of the genetic potential of the pig to make it nearly human, it could grow properly any more in a pig. Moreover, there was no guarantee that pig organs could do the job of a human organ for long enough to make xenotransplantation worthwhile.

A bigger issue than rejection was the fear that the use of pig organs could create new diseases. In particular, scientists such as Robin Weiss of University College London worried about a group of viruses called porcine endogenous retroviruses, or PERVS. The DNA genetic codes of the viruses have over millions of years become an integral part of the pig's genetic code. These "duplicate me" DNA messages are harmless to pigs, and not easy to spread, but critics argued that no one knew what would happen if they were introduced into humans in transplanted organs. As Weiss put it, "Beware the Trojan Pig! When taken into your body, it may release viruses hidden inside its organs just like the Greeks in the belly of the horse."

Intriguingly, my colleagues at the Roslin are now looking at how to introduce alpha gal into human stem cells. This is not to make them more porcine but, paradoxically, to increase the success of their transplant into patients. One fear is that if nerves differentiated from the cells are, for example, transplanted into the brain to treat Parkinson's disease, they may still contain embryonic cells capable of developing in an unregulated way into a teratoma tumor (as with any therapy, there are dangers as well as opportunities). My Roslin colleague Jim McWhir has studied how to use genetic modification to ensure that alpha gal is turned *off* in differentiated cells but turned *on* in the embryonic cells. That way the embryonic cells would be seen as foreign and rejected if accidentally grafted into a patient.

COWS, JAPANESE BEER, AND HUMAN BLOOD

THE GOING is always tough for any scientific pioneers who try to be the first to translate their bright ideas from the laboratory into the marketplace. Nexia Technologies in Montreal wanted to milk fast-maturing dwarf transgenic goats for orb spider silk proteins. Two transgenic kids, Peter and Webster, were born in 2000 with the spider silk gene incorporated into their genetic makeup, the founders of a herd that the company hoped would produce a synthetic form of spider silk called BioSteel. Bulk spider silk could be used for all sorts of applications, from body armor to medical sutures. But in May 2003 the company ran into trouble, laying off employees to save money because of delays in developing a commercial spinning process. Using the same approach, the company moved on to goats whose milk contains human butyrylcholinesterase, a protein that seeks out and binds to nerve agents such as sarin and VX.

PPL Therapeutics, the best-known of the Roslin spin-out companies, sold its Virginia subsidiary to the University of Pittsburgh in 2003 to stay afloat (in 2004 PPL had left the stock market altogether with a value of just £7 million, compared with around £500 million in the heady days when Dolly had put the company on the front pages of newspapers across the planet). PPL's xenotransplantation technology was the most advanced of what it had to offer Pittsburgh, but it had also been attempting to create an "infectious disease platform technology." What it meant by this was the use of GM animals to make human antibodies that are normally found in human blood, where they help to fight infection. Most therapeutic antibodies already on the market consist of a collection of identical (monoclonal) antibodies. But a real prize would be to make a collection of many different types of human antibodies (polyclonal antibodies), which would have many different applications, and to do so by means of animals rather than people.

In recent years Hematech and its collaborator, the Tokyo-based Kirin Brewery Company, have managed to do just this by combining cloning and genetic modification to alter cows—a potentially great source of these precious proteins since a typical mature cow has about 2.5 pounds of antibodies in its blood. Hematech in Sioux Falls, South Dakota, has unveiled a plan to develop the "Tc cow," which it claims would be able to make human polyclonal antibodies that can help fight antibiotic-resistant infections, fend off bioweapons such as botulinum toxin and anthrax, and treat immune deficiencies and other illnesses.

To create the Tc cow, the team, led by Chief Scientific Officer Jim Robl, has had to overcome several major technical hurdles. The first was how to handle the genetic wherewithal to make polyclonal antibodies. The region of chromosome that carries all of the genes that direct the production of human antibodies is huge, far bigger than any other sequences that have been transferred into livestock. To make matters more complex, it is hard to grow cells from livestock in culture for very long, so it is harder to carry out genetic modification and check that modification has worked than, say, with mouse embryonic stem cells, which grow almost indefinitely.

Jim Robl came up with two clever strategies to sidestep these difficulties. First his team assembled an artificial chromosome to carry the heavy burden of DNA required for transfer of the antibody genes. Added to the chromosome were DNA sequences that offered protection against a drug. This artificial chromosome was then introduced into fibroblasts grown from a calf in the presence of the drug. Only fibroblasts that took up the protective DNA, and presumably the new chromosome, survived. However, rather than carry out a large number of time-consuming tests to confirm that the artificial chromosome was being used, the team adopted a second innovation—to use the cells for nuclear transfer as early as possible. It allowed the clones to develop for fifty days before they were recovered and cells grown in culture. This offered two advantages. By the process of cloning the nucleus was rejuvenated so that the cells had a longer life span in culture, and as a result it was possible to discover whether the cells had the additional chromosome. The cells where the alteration had taken hold were then used for a second series of nuclear transfers, which produced the transgenic calves.

But these "transchromic" calves were still not ready for pharming. They produced both human and bovine antibodies that would be difficult to separate for clinical use. To overcome this, Robl and his team wanted to use a process called sequential gene targeting, by which the team could knock out both copies of the bovine immunoglobulin gene. An additional hope was that cutting the bovine antibodies circulating in the blood of the animal would significantly increase the yield of a human antibody collection (polyclonal antibody). That work is under way as I write. Robl and colleagues hope that by immunizing the Tc cow with a disease agent, they can create polyclonal antibodies in the bloodstream that have a strong reaction against that agent. Hematech aims to harvest antibodies from Tc cows on its production farms—around 125 head each—using a process called plasmapheresis, in which whole blood is collected, the cells are separated from the liquid part of the blood (plasma), and the cells are returned to the body. After being collected, the plasma would be shipped to a purification facility where the cow blood components (proteins, fats, potential bacteria, and viruses) would be removed. The sterile, purified antibodies would then be formulated into the final product shipped to hospitals. In 2004 Hematech and Kirin also announced they had genetically engineered a bovine embryo by removing both copies of the prion gene. The hope here, as with our earlier work, was that animals produced from these embryos would be unlikely to contract mad cow disease. And the animals would be ideal for the production of the companies' fully human polyclonal antibodies. Pharming is now under way, not of milk or beef but of human antibodies.

Goats are being used to make another promising pharm drug, called ATryn, a human antithrombin mentioned earlier. The drug has been developed by GTC Biotherapeutics of Framingham, Massachusetts, to treat hereditary antithrombin deficiency, a condition that makes patients vulnerable to deep-vein thrombosis—damaging blood clots. This therapeutic protein is derived from the milk of transgenic goats, selected because they are cheap, easy to look after, quicker to mature than cows, and able to produce up to three kilograms of human protein annually. To create the goats, the scientists inserted the human gene for the protein into a fertilized goat's egg and, to ensure that it is activated only in the mammary gland, and efficiently, added extra DNA: regulatory regions from the gene of a milk-specific protein, other DNA regions that help assure high-level protein production, and a piece of backbone DNA that allows the construct to be handled, replicated, and microinjected into embryos. In this way the goats were carefully bred to maximize milk production, and thus drug production, by GTC Biotherapeutics. The company's boss, Geoffrey Cox, says his firm has created sixty-five potentially therapeutic proteins in the milk of its transgenic goats and cows. Given the longer than expected gestation time of this technology, the possibility of unforeseen problems, and the exceedingly slow pace at which regulatory agencies move, it is anyone's guess when society will notice the benefits.

GENE WIZARDRY

MANY ADVANCES in the efforts to make it easier and more efficient to genetically alter animals have been made in parallel to these developments. One of the most important is a new and highly precise way to toss a spanner into the works of a cell to shut down, or "silence," genes. This advance rests on new insights into genetic code called ribonucleic acid (RNA), long thought merely to be the dowdy ancestor of its glamorous cousin deoxyribonucleic acid (DNA). Like DNA, RNA consists of strings of chemicals—"letters"—that spell out the genetic code. The letters are represented by A, C, G, and either T in DNA or U in RNA. C always binds to G. A binds only to T or U. In this way, a single strand of DNA or RNA can bind to another strand consisting of complementary letters, the same feature exploited by DNA targeting.

Until recently, it was thought that RNA simply carries out DNA's commands, by translating genetic information into action. Then the DNA's double helix unwinds, and its genetic code is copied onto a single-stranded "messenger" RNA. In turn, the messenger RNA shuttles the code from the heart of cells to ribosomes, the "factories" that make the proteins that build and operate cells. But there is more to RNA than this. The first hint of RNA's versatility came some years ago, when Andrew Fire of the Carnegie Institution, in Washington, D.C., Craig Mello of the University of Massachusetts, and colleagues found that double-stranded RNA can be used to shut down specific genes. Their discovery, called RNA interference, or RNAi, is now widely used as a research tool. Another clue came from David Baulcombe and Andrew Hamilton in the Sainsbury Laboratory, in Norwich, England. They discovered RNA fragments called short interfering RNAs, or siRNAs, which cells use to fight viruses. Although they were working with plants, their work complemented the animal studies of Fire and Mello because siRNAs are fragments of double-stranded RNA.

RNA interference seems to occur when double-stranded RNA is chopped into siRNA. The siRNA then guides an enzyme that degrades the messenger RNA that turns each gene into a protein. This technique caused a lot of excitement because gene silencing is highly specific—the messenger RNA will be degraded only if it has the same sequence of A, C, G, and U letters as the double-stranded RNA. This means a messenger RNA—for example, from a disease gene—can be targeted without the risk of side effects that would occur if the other messenger RNAs needed by a healthy cell were silenced. At Cold Spring Harbor Laboratory, on Long Island, researchers have also tried using RNAs folded over like hairpins to quash specific genes. These seem to work as efficiently as siRNAs, heralding what they playfully call short hairpin activated gene silencing, or SHAGging. The use of RNA interference is growing rapidly and offers another useful tool for pharmers.

DOWN ON THE PHARM

WE HAVE COME a long way since we tried to combine cloning and genetic modification to make the protein that Stephen Christmas lacked. Our efforts were, unfortunately, much too late to help him, though he was as influential at the end of his life as at the start, when he first gave his name to a form of hemophilia. Stephen worked for the Canadian Hemophilia Society and campaigned for the cause of blood safety, helping to secure financial compensation for patients infected with human immunodeficiency virus, HIV, despite the foot-dragging by the bureaucrats and authorities.

The plasma from as many as 100,000 donors was combined to make blood products, such as the factor IX concentrates used to treat Stephen's hemophilia B. In the early 1980s factor concentrates were not heat treated, because this made them less effective. In those days all it took was one HIV-infected donor to contaminate the lot. As a result roughly 80 percent of all treated hemophilia A patients and 50 percent of all treated hemophilia B patients became infected with HIV-1.

A Canadian Hemophilia Society report, prepared largely with archival documents uncovered by Stephen Christmas, prompted the federal government to provide victims of the tainted-blood tragedy with compensation. Stephen learned that he was likely infected while sifting through recall notices for blood products in 1987, two years after the testing of blood donations for HIV and heat treatment of blood products became mandatory in Canada. He died five days before Christmas Day in 1993, aged forty-six. His older brother, Robin, remarked on the irony that Stephen had been killed by the factor IX that had initially given him "a shot at life." By then Robin had married a Jewish woman, Linda Rosenbaum, but they always celebrated Christmas in the traditional English way (Linda even learned how to make plum pudding) since they were Stephen's only close family. "When Stephen died, so did our Christmases," she said. "But we always make sure to especially talk about him on that day. And we light the Jewish memorial candle of remembrance every year on the day of his death."

In his final years Stephen had been bitter and angry about the authorities' indifference to the plight of infected hemophiliacs. As if to reinforce this, his death came before administrative paperwork had been completed, so his estate missed out on some compensation. But his friends and family described his lifelong battle with pain as a journey of courage, learning, and peace. "He was an upbeat, well-read, and well-spoken gentleman," said one friend and fellow hemophiliac.

Our dream of using a single flock to provide the world with a safe supply of factor IX was not realized as quickly as we had hoped in the optimistic days when Polly was born. Perhaps we should have known that, like any radical new technology, it would take longer to become established than we had expected. Striking the right balance between commercial hype and either social or scientific conservatism is always difficult. In the case of genetically altering animals so that their organs are suitable for xenotransplants, an impressive alternative has emerged from the very technology that helped make it all possible: nuclear transfer for therapeutic cloning.

Rather than be used for efforts to "humanize" pig hearts, stem cells offer a way to grow a patient's own heart muscle to carry out repairs. I remain an enthusiastic proponent of using cloning to carry out genetic alteration more accurately and safely than ever before— for instance, to make animals resistant to disease. However, much more excitement has been generated in recent years by stem cell science. I myself have given up working in muddy fields and doing surgery in operating theaters for cloning human blastocysts and growing lawns of human stem cells in the laboratory.

FIVE

CLONING FOR
MY FATHER

MY FATHER, JACK, was just one person among millions who might have been helped by cells derived from a human embryo. He was diagnosed as being diabetic when he was just twenty-two, while finishing his mathematics degree at Magdalene College, Cambridge. He was gifted and, on graduation in 1935, was offered the chance to do research. But he had other ideas. He was passionate about teaching and went about inspiring the next generation with his enthusiasm for mathematics.

Jack Wilmut was not a big man, at five feet four and 147 pounds, but had no problem with school discipline. He had great presence and was highly effective in the classroom. I admired him immensely. Perhaps I resemble him a bit in character. I would like to think so. People often seem to regard me as a schoolteacher. And I do enjoy giving lectures at conferences, meetings, and public events.

My father managed to teach for only two decades, despite his ability and drive. Gradually, diabetes robbed him of his sight. His vision deteriorated to the point that it was impossible for him to teach in school. By the time he was in his forties, he was blind. Remarkably, he bore his suffering patiently and wanted to reinvent himself, this time as a computer programmer, a leading-edge profession of the day.

At that time the government offered retraining for another career only to younger men. My father wrote to the current prime minister, Harold Wilson. He was emboldened by the thought that the PM owed a personal debt to the Wilmut clan: Wilson had been given remedial lessons by my grandfather, Sam, after suffering rheumatic fever as a boy and being away from school for a prolonged period. This fact Wilson appreciated and never forgot. The employment rule was changed, and my father was able to learn a new skill, one better suited to his failing body. I would like to believe that this change happened because of Wilson's response to my father's appeal.

My father went on to work as a computer programmer for Yorkshire Imperial Metals in Leeds. But the disease continued to attack his body. He gradually lost the sensitivity in his fingertips needed for Braille. Once again he was forced out of a job. He had to retire early. Before he died, in 1994, he lost much of the use of his fingers and part of a leg was amputated.

Like any son or daughter, I would have done anything to reduce his suffering and make him better. If only there had been a way to halt the underlying disease by replenishing my father's body with his own islet cells, the insulin-producing cells that he lacked. One day it will be possible to do just this, by means of an embryo. From this embryo could be isolated stem cells, which have the potential to turn into any of the two hundred different types in the body.

The embryonic cells could be allowed to differentiate and, from the muddle of resulting cell types, the right ones be selected to fix an ailing body. Better still, embryonic cells could be guided with growth factors down the right developmental path to make desired kinds of cells, offering huge opportunities to repair the effects of illness, accident, or disease. For my father's disease, these precious embryonic stem cells could be turned into islet cells. Aside from providing cells for transplant, and means of testing diabetes drugs, this research will also yield insights into the development of the disease. And, of course, Dolly is the spiritual mother of all this work since she showed unequivocally that differentiated adult cells could be reprogrammed to be embryonic again.

The usual term for the procedure of deriving cells from cloned embryos, "therapeutic cloning," sends shivers down the spine of many people. Many experts in science and ethics, such as Mary Warnock, the philosopher who chaired the UK's Committee of Inquiry into Human Fertilisation and Embryology have wondered whether the use of the term "cloning" has damaged the field because it is so laden with grim associations and negative baggage. Understandably, the official alternative—"cell nuclear replacement"—is gray and wordy and, as a result, has not caught on. What's in a name? In this case, a great deal. These primal cells are the stuff of which medical dreams are made.

Embryonic stem cells offer the potential for making not only insulin-producing cells to treat diabetes but also nerve cells to treat degenerative diseases like Parkinson's, heart cells (called cardiomyocytes) to treat a damaged heart, and other cells to fix liver damage from hepatitis or alcohol abuse. They could be turned into twitching heart cells to test new cardiac drugs, or nerve cells to rewire a broken spinal cord—you name it. In fact, so-called cell therapy is already well established. We are all used to the idea of bone marrow transplants to treat leukemia and also understand that this established treatment fails if donor marrow is not a good match and the graft attacks its host rather than protecting it: stem cells offer a way dependably to obtain genetically compatible tissue of any kind.

Initially, these cells have been obtained from "spare" embryos from fertility treatments, and it is estimated a bank of a few dozen different types could be of practical use. Even if a stem cell line produced a good match, however, it would still be immunologically different from the patient. Transplant recipients would have to take powerful and toxic antirejection drugs for life—drugs that increase susceptibility to infection and even cancer.

By using cloning—nuclear transfer—to grow a patient's own tissue, one can avoid rejection (there is a complication when it comes to treating the diabetes that affected my father, however, which I will discuss later). Stem cells from a blastocyst have almost, but not quite, all the potential of the newly fertilized egg. After starting out as the single-cell fertilized egg, the embryo divides into three different layers—ectoderm, mesoderm, and endoderm. The ectoderm turns into nerves, skin, and hair follicles; the mesoderm produces heart, blood, and muscles; and the endoderm becomes the organs of the intestine, liver, pancreas, and lungs. In a similar way the embryonic stem cells can turn into any of these cells.

Even in a mature adult there will always be some more stem cells, which are at a later stage of development than embryonic stem cells but an earlier stage of differentiation than liver, brain, bone, and so on. Another option would be to use these more developed kinds of stem cells, adult stem cells, which can develop into a smaller repertoire of cell types for local replenishment and repair of an organ. Their flexibility is still under investigation. Whereas stem cells from adult bone marrow, for example, are known to turn into blood cells, great excitement was generated by experiments that seemed to suggest they were much more flexible, generating brain cells, liver cells, and other types. The pro-life movement has placed great faith in these observations, claiming that they undermine the case for using embryonic stem cells. I too would welcome any use of adult stem cells, but questions remain whether the range of cells that they can be turned into is as broad as that possible with embryonic stem cells.

Austin Smith, while at the Institute of Stem Cell Research, Edinburgh University, reported in 2002 that some adult stem cells can fuse with existing cells, creating freak cells with double the dose of chromosomes—eighty instead of the usual forty—and unknown health consequences. In the same year Catherine Verfaillie of the University of Minnesota Stem Cell Institute found that cells from adult bone marrow could provide an alternative to embryo cells to make brain, heart, muscle, or any other type of cells. But she herself emphasized that it was too early to say how bone marrow cells would compare with embryonic stem cells in longevity and function. It was already known that stem cells from marrow can make bone, cartilage, and muscle, but the question was, and remains, whether these other cells can make anything beyond this expected range.

The pro-life movement still argues that there is evidence that adult stem cells, such as those from bone marrow, can "transdifferentiate" and be turned into nerve, heart, muscle, and other types of cells for repair, and that this makes human embryo research redundant. However, Austin Smith remains skeptical and believes that many claims to this effect have come from overenthusiastic researchers in countries where human embryonic stem cell research is severely limited. Even how to define a stem cell, let alone its potential, remains hazy at the time of writing: "There are useful stem cell types in the adult," says Smith, "but the proposed paradigm shift, that blood can make brain and vice versa, has not stood up to the scientific test."

It has been claimed that adult stem cells from bone marrow—among a wide mixture of cells—can repair a heart, and success has even been reported in patients with heart disease. But Smith argues that it is unclear that new blood vessels and heart muscle cells are arising from adult stem cells injected into a diseased heart. Instead, the benefit may come from growth factors and other proteins exuded by the transplanted cells, which have stimulated heart tissue repair. Not everyone is so gloomy, however.

CELL BANKS

TO DETERMINE which stem cells are most useful in treating a particular disease, side-by-side comparison of adult and embryonic varieties of stem cells must be done. This is one of the reasons that Britain launched a stem cell bank in 2004, the first of its kind in the world, at the National Institute for Biological Standards and Control, in South

Mimms, where ampoules of human stem cells of various types are laid down for storage in liquid nitrogen.

For those who regard a fertilized egg as a person, the very idea of the bank must be unsettling. Within its freezers the genetic identity of these "individuals" can live on almost indefinitely. The nature of the cells raised "distinct problems," according to a House of Lords select committee on stem cells. There are already many examples of "immortalized" cell lines used in research where these issues have emerged. Perhaps the best-known example was created more than half a century ago. On 4 October 1951 Henrietta Lacks died of cervical cancer at Johns Hopkins University Hospital, in Baltimore. The thirty-one-year-old mother of five might have died in obscurity, had it not been for her cancer cells, now known as HeLa cells, which were the first discovered to thrive and multiply outside the body, seemingly forever. Her cells have been used in endless experiments and even been launched into space.

Critically, for stem cells to be stored this way there should be no restrictions placed on their use by the parents of the embryo from which they came. In recognition of this and pro-life concerns, the bank was established only after a long haul that involved mulling over many legal and ethical issues. While researchers were most interested in how to reliably grow stem cells into brain, nerve, muscle, or any other desired type of cells, they also had to ensure that cells were grown in conditions of "good manufacturing practice"—essentially keeping a record of anything and everything that goes on in stem cell facilities—so that a convincing case could be made for the cells to be used to treat patients. Use of the cells is heavily regulated: in addition to the UK's Human Fertilisation and Embryology Authority, the Medicines and Healthcare Products Regulatory Agency has set conditions that have to be satisfied. The owners of any patents on processes used to create the cell line also control how the cells can be used.

Cell lines from around the world are now being sent for storage in South Mimms. For the bank those cells could be adult stem cells, such as bone marrow cells, those from a fetus aborted for medical reasons, or those from a cloned or surplus embryo, such as a blasto-cyst. As I have said before, the early embryo deserves respect but so do the aims of the bankers. These precious cells will accelerate research in a field that offers new hope for a range of currently untreatable conditions. The bank will also make the most efficient use of a controversial source of tissue: no one wants to make or use any more embryos than necessary. In addition to making every embryo count, the bank is critical in helping to break a monopoly of private companies and to assist the best scientists to work on under-standing these cells.

Once warmed and grown at blood temperature in nutrients and growth factors, the stored stem cells can be used to find ways to reli-ably turn them into the various cell types in a human body. After the science is perfected, a single cell can fill freezers with ready-to-use immune, brain, heart, or liver cells. Turning them into tissues will be trickier, and then turning them into organs remains a distant prospect—how can one nurture them without the complex, evolving environment of a body that grows in tandem?—but a ready supply of replacement cells alone could revolutionize the treatment of Alzheimer's and Parkinson's diseases, stroke, spinal cord injuries, heart disease, and diabetes.

A good match for most people could come from a bank of just 150 lines of embryonic stem cells, according to studies by Roger Ped-ersen's team in Cambridge, England. Pedersen had left the University of California, San Francisco, to take advantage of an attitude toward embryonic stem cell research that is more liberal in the UK than in America. In the UK he found government support for such work (at the time he said he faced a choice between "very favorable circum-stances and tremendous support" for his work in the UK and "the prospect of sitting on [his] hands for the next few years" in the States). Working in Cambridge, his team found that as few as ten carefully selected lines would be sufficient to meet the needs of 40 percent of the population, with the aid of antirejection drugs.

THE GREAT CELL HUNT

REPORTS IN THE MEDIA describing stem cell research have created the impression that the feat of isolating human embryonic stem cells is trivial and unremarkable. It is not. The first success was reported in 1998, after six years of trying and almost two decades after a British team first carried out the feat in mice. The first line of human embryonic stem cells was derived by James Thomson and his team at the University of Wisconsin, Madison. In November of that year, when the feat was described by his group in the journal *Science*, Thomson said, "It's no longer in the realm of science fiction. I really believe that within my lifetime I will see diseases treated by these therapies."

The cells were licensed to the Geron Corporation in California, evidence that even then there was intense commercial interest in their clinical uses. Only a few days later John Gearhart of Johns Hopkins University reported success in cultivating a line of stem cells from the germ cells of aborted fetuses. "The potential of these unique, versatile cells for human biologic studies and medicine is enormous," said Gearhart. "These cells will rapidly let us study human processes in a way we couldn't before. Instead of having to rely on mice or other substitutes for human tissues, we'll have a unique resource that we can start applying to medicine."

Human embryonic stem cells have rarely been out of the headlines since, and the center of gravity of the research has shifted from the West to the East, with teams in Israel, China, and Singapore making great strides. But in Britain it took researchers five years to catch up with the initial advances made in North America. Isolating a line of stem cells proved extraordinarily difficult: researchers needed not only "green fingers" to succeed but also to guard against infection, contamination, or colonies morphing into mixtures of other tissue.

Like many other groups, we study Thomson's original lines of cells. Like many other groups in the UK, the group at the Roslin was also keen to isolate the first fully characterized line of human embryonic stem cells in the country. Various other groups wanted to have a go as well; in Edinburgh led by Austin Smith; in Sheffield led by Peter Andrews and Harry Moore; Miodrag Stojkovic's, Majlinda Lako's, and Alison Murdoch's team at the Centre for Life, Newcastle; and one led by Stephen Minger—another American—and Susan Pickering in King's College in London.

The King's team was the first to report success, in the Internet journal *Reproductive BioMedicine Online*. Its paper described how the King's stem cells made two molecules unique to human embryonic stem cells, as well other genes commonly found in other stem cells. This was suggestive, though not hard, evidence that they had the elusive cells. But the King's team felt that its work warranted publication because it used embryos that had failed screening tests for serious genetic disease (by a method called preimplantation genetic diagnosis, where a cell is removed from an early embryo for analysis, which I discuss in chapter 8). There is always a fear that, if the embryos are of poor quality in one respect or another, it may be impossible to derive a line of stem cells. Thus the team members considered it worth announcing that they had helped open up a source of stem cells that (at least to pragmatists) are more ethical to use.

The race to be the first in the UK probably ended in a dead heat, however. The Newcastle team had been growing its stem cells since the start of 2003 but delayed publishing results so that it could implant the cells into mice that lack an immune system to see whether the cells turn into teratocarcinomas, cancers that consist of a blend of all cell types. This experiment was regarded as a crucial test to show that the cells are "pluripotent," that is, capable of turning into any type. The Newcastle team, which derived the stem cells from a more conventional source—embryos left over from IVF treatments—did not feel it could publish until this test had been done. The danger of a premature scramble for publicity was emphasized by an earlier false dawn in British stem cell science. In August of the year before, an overenthusiastic BBC news program reported that King's had pulled off the feat. That cell line was subsequently lost to contamination.

ROSLIN'S STEM CELLS

THE ROSLIN'S stem cell effort began in earnest in May 1999, when the institute signed a deal for research on human embryo stem cells with the Californian company Geron. As part of this program we set up a "clinical grade" laboratory for the isolation and growth of stem cells, led by Paul De Sousa and backed by Scottish Enterprise and the Medical Research Council. Like other teams, we were excited by the thought of answering a very simple question: if nature can direct a single, nonspecialist übercell—the embryonic stem cell—to develop into masses of specialist cells, such as bone, heart, muscle, and so on, why can't we?

We wanted to extract the essence of "stemness" and discover what allows stem cells to specialize to become a member of a tribe, whether the skin tribe or the eye tribe. Several factors were known to influence how to turn a stem cell into an endoderm cell, a cell that could help rebuild the pancreas of a diabetic patient, such as my father. The challenge facing my team and all the others was, and remains, finding ways to decipher all the components of the complex chemical environment within a body that reliably determine a stem cell's fate.

In all, it is estimated, there must be about twenty pathways in the body governed by hormones and growth factors that control how a stem cell turns into a particular cell type. Like other teams, we are looking at how to tweak all twenty in a systematic way by the use of robots and automated methods. For example, it is possible to put the cells into trays of tiny wells, some ninety-six in all, and then use robots to move between the wells and alter conditions one by one so that experiments can be conducted on all of them simultaneously. One could thus have a gradient of concentrations (high to low) in two different chemicals in the recipe across the vertical and horizontal wells of the microdish, so one can test the influence of both on the ease with which one can incubate, grow, and differentiate the cells. By using 100 plates, you can look at 10,000 conditions. In a few weeks you can complete a set of experiments that would keep a conventional laboratory tied up for a year. As a bonus, these microdishes work with a minimal number of cells, growth factors, and other precious substances (human proteins cost a fortune).

By finding a precise recipe that can reliably turn stem cells into the desired type, we also cut the risk of infection by animal or human diseases. The reason is that the traditional way to grow human embryo stem cells is to nurture them on animal (usually mouse) "feeder" cells in a broth that is as unscientific as it sounds, consisting of fetal calf serum, a complex mix of substances with an action that is not fully understood.

In 2005 the risks of this approach were underlined when contamination was found in the "presidential" lines of human embryonic stem cells—President George W. Bush had in 2001 restricted federal funding to develop new treatments to the limited numbers of embryonic stem cell lines available at that time, a move that annoyed pragmatists and fundamentalists alike. Ajit Varki, from the University of California, San Diego, and Fred Gage, of the Salk Institute, found that the lines contain a foreign substance, called N-glycolylneuraminic acid (Neu5Gc), and suggested that if they were put into patients there was a significant chance of a deleterious immune reaction and/or rejection of the transplanted cells.

As one way past this, new cell lines being derived at the Roslin by Paul De Sousa, backed by Geron, have been grown with the help of fibroblasts from human foreskin and human laminin, from placentas, but the hope is that one day we can do the equivalent for human cells of hydroponics for plants, where soil is replaced by a solution of nutrients. We wanted to get rid of these feeder cells altogether so that we knew exactly what our stem cells were dining on and how to ensure that this recipe produced cells of the desired type. We have to get this recipe absolutely right if these cells are to be safe to use.

If a stem cell runs amok, the special type of cancer called a teratocarcinoma can result, which is a grotesque tribute to the stem cell's flexibility. The cancer is a mixture of gut, muscle, nerves, teeth, tufts of hair—even facsimiles of fingers and legs—that I mentioned earlier as one of the tests of pluripotence. We have to make sure that the cure is not worse than the disease.

Eventually, with colleagues at the University of Edinburgh, Jim McWhir hopes to turn stem cells into bone cells and cartilage, while Wei Cui's team is studying how to make liver cells for drug testing (since the liver enzyme cytochrome p450 breaks down most drugs) and the right kind of brain cells to treat Parkinson's disease. As well as using established lines of human stem cells, my colleagues are trying to create new lines in these defined culture conditions. One source of blastocysts is IVF, when they are surplus to requirements or have failed screens for congenital disease by preimplantation genetic diagnosis. In both cases these embryos would otherwise have been discarded. Instead, and with the permission of the parents, they can be used for experiments. Most people would accept that this is an ethical source of cells, particularly given the usual fate of these blastocysts.

VIRGIN CELLS

THERE IS, however, an alternative source of human embryonic stem cells that does not—or at least *should* not—count as a human life. The human embryo in question results not from a fertilized egg but from an egg that has been tricked into dividing. It is as if it were possible, for a short while, to rekindle some of the ancient chemical machinery that enabled single-cell creatures to clone themselves.

Paul De Sousa and I have been given permission to realize an idea that dates back to the start of life on Earth and gets its name from the ancient Greeks; "parthenogenesis" literally means "virgin birth." One early reference to parthenogenesis can be found in Greek mythology, when Athena emerged from the forehead of Zeus (Athena's temple on the Acropolis is called the Parthenon for this very reason).

Most plants can multiply by parthenogenesis, and so do all fungi and many animals, such as corals. Other examples range from aphids and bees, the male drones of which come from unfertilized eggs, to the timber rattlesnake and the basilisk, called the "Jesus Christ lizard" for its ability to walk on water. Even the turkey can do it.

But parthenogenesis cannot naturally give rise to offspring in mammals, such as humans. The reason has to do with a process called imprinting, in which the mother and father have a say in how genes are used. Any attempt to persuade an unfertilized mammalian egg to develop to term is doomed by imprinting, a phenomenon I will explain in detail in chapter 7. Briefly, although you inherit two copies of every gene, one from each parent, only one is used in the body. Imprinting governs which one—whether the version from the mother or the version from the father. In this way, mums and dads shape how their genetic inheritance is spent. In mammals at least seventy-five genes (probably many more) with diverse and important functions during development are thought to be regulated in this way, so they must be inherited from the right parent for a pregnancy to succeed.

Imprints are critical for the creation of the placenta, the embryo's lifeline to the mother, and disruption of imprinting of other genes is linked with schizophrenia, immune deficiency, and forms of cancer. Imprinting also explains the cause of Prader-Willi and Angelman syndromes, marked by mental retardation and a host of problems: they strike when a baby fails to inherit a paternal copy of chromosome 15 or fails to inherit a maternal copy, respectively. And imprinting explains why there are no virgin births: the embryo cannot develop without a genetic contribution from the father. At least not to term.

THE MOUSE WITH
TWO MOTHERS

WE STARTED OUR parthenogenesis effort at the Roslin before the birth of a remarkable rodent that highlights the importance of imprinting, how it arose and also how to defy it. Joining Dolly the sheep on the list of headline-grabbing creatures that have pushed forward the frontiers of science, we must add Kaguya the mouse. Although she supposedly came into this world through parthenogenesis, at least according to the journal that announced the feat, scientists have argued about what to call a mouse that has two mothers, rather than just one.

Kaguya leapt over the apparently huge obstacle of imprinting that prevents a mammal parthenogenetic embryo—a "parthenote"—from developing. She was born in 2004 as a result of an effort by Tomohiro Kono of the Tokyo University of Agriculture, with colleagues in Japan and Korea. Rather than combine the imprinted male and female genetic codes, as normally happens after egg meets sperm, Kono combined a female genetic code from an older egg with a "male-like" female genetic code created from the genetic makeup of a young egg, which is free of most imprinting.

To masculinize the "imprint-free" female code, he used genetic engineering to remove a gene (called H19) from the young egg so that it was being used only in the old egg. As a result of the removal of this gene, production started of a growth factor (Igf2) that is normally made only from the male genetic code. The factor plays a key role in the development of the placenta, among other tissues. The first mammal produced with two mothers, named Kaguya after a Japanese fairy-tale princess, grew to be a healthy adult female that was able to produce offspring the old-fashioned way.

Kaguya provides support for a theory that explains why the peculiar business of imprinting evolved in mammals. Put forward by David Haig of Harvard University and colleagues, it rests on the most fundamental feature of all life: the urge to propagate one's genes to future generations. Imprinting results from an evolutionary battle between the sexes, reflecting the different genetic agenda of each parent, according to Haig. To put it crudely, both parents want to pass on their genes to their offspring while still fulfilling the selfish urge to "duplicate me." Conflict arises because a woman's genes also have her interests—not the man's—at heart, and vice versa.

The father wants his offspring to grow big. This will provide his offspring—and his genes—with an enhanced chance of survival and use up the mother's resources so that they won't be wasted on another man's offspring. Women, on the other hand, want to share resources with the unborn child and yet reduce the risk posed by unlimited fetal growth, which would be deadly. The mother also wants to survive to pass on her genes to future children.

The result is an evolutionary arm-wrestling match, with paternal genes beefing up the offspring and maternal genes holding growth in check. Thus parthenogenetic human and mouse embryos fail to develop because they contain a double copy of growth-limiting maternal genes without paternal rivals such as Igf2 to boost growth. Kaguya overcame this obstacle. But it is still a riddle why the Tokyo team's simple genetic modification managed to jump this hurdle, albeit in only 1 out of 450 or so attempts.

In our experiments we were not trying to make a human Kaguya. We thought that human parthenotes could aid our work on therapeutic cloning while sidestepping some of the ethical issues. And we did not have to use genetic modification, because imprints seem to have the most influence later in development, making it possible to defy them for a few days. A human egg can be persuaded to divide with a shock of electricity or dose of chemicals. The unfertilized egg can also be tricked by chemicals into keeping the set of chromosomes that it would have otherwise have jettisoned in polar bodies. (In normal sexual reproduction, an egg loses half of its chromosomes, so when it docks with sperm during fertilization the full number is restored in the resulting embryo.)

The result is a parthenote that retains the full set of DNA of one woman. Although it cannot develop to form offspring, indeed does not develop for more than a few days, the parthenote can provide insights into imprinting and provide a source of embryos with which to learn how to derive stem cells. It is uncertain whether such cells could be used to treat patients because of the errors in imprinting. However, if these errors do not interfere, cells from parthenotes could be used to treat the woman who had donated the egg for parthenogenesis. During the lengthy process of egg development (meiosis) the genes of a woman's mother and father are scrambled by a process called recombination, so her eggs have a different genetic makeup. Still, they should not "look" foreign to the egg donor's immune system, and the cells could be of value for treatments where an exact tissue match is probably not required, such as for the creation of nerve cells for transplant into the "immune privileged" brain to treat Parkinson's. Parthenogenesis also offers the opportunity to create a cell line from a woman who is a sufferer of a serious genetic disease, so that a very detailed study can be made of the cellular effects of the disease.

The parthenogenesis program also helped with the most controversial part of our embryonic stem cell research at the Roslin: the use of nuclear transfer to create embryos. Part of the nuclear transfer method involves using electricity or chemicals to persuade a reconstructed egg to divide, and here the methods—"activation protocol"—used to trick an egg into dividing to form a parthenote provide vital information on the best way to do this and to launch development.

NEVER SAY NEVER AGAIN

THE SAME YEAR we were granted a license to create human parthenotes, some scientists began to suspect that humans might be particularly hard to clone. They had, of course, forgotten the central lesson of recent reproductive science: never say never.

In 2003 Gerald Schatten, director of the Pittsburgh Development Center, observed there were "molecular obstacles" that blocked normal cell development in cloned rhesus macaque embryos. His team tried four different cloning methods on 724 eggs retrieved from female rhesus macaques. In the resulting embryos the chromosomes did not separate properly when cells divided.

A fundamental process in cell division is the pulling apart and segregation of chromosomes equally into the two daughter nuclei; to do this, the cell draws on the cytoplasm to build a threadlike structure called the spindle to precisely align and separate them. An abnormality in the spindles of human embryos before implantation, because of abnormal replication of centrosomes (the organizing centers and precursors of spindles in the cell), is thought to be one reason for many of the chromosome defects observed in early human development.

As a result of cloning, however, Schatten found "chaotic" spindle structures and unequal chromosome counts. Even the most basic proteins involved in spindle formation were absent or inadequate. Schatten concluded that when it came to primates and people, therapeutic cloning would be difficult and "reproductive cloning"—the creation of cloned babies—almost impossible. Even at the time, this seemed too pessimistic. Perhaps, in nonhuman primates, the normal procedures for removing the genetic information from the egg (enucleation) also extruded something important from the cytoplasm. Perhaps a protein that plays a key part in the formation of the spindle was taken away too, wrecking the prospects for development. As ever, small changes to the basic cloning method could overcome this. For example, the donor nucleus could be put into the egg before removing the egg's genes. An article in the journal *Science* concluded, "There are potential ways around the newfound obstacle, but for now, groups that made controversial claims that they would use the techniques that produced Dolly the sheep to create human babies are unlikely to succeed."

Given the pressures on the American scientific community from the Christian Right, one can't help wondering whether some scientists yearned for a technical roadblock that could somehow end all hopes of reproductive cloning and yet keep the vast promise of therapeutic cloning alive. The echoes of Solter's gloomy pronouncement on the prospects of cloning mammals were unmistakable. There were also more distant echoes of that age-old human desire to feel special, unique, and somehow separate from sheep, amphibians, and the rest of the animal kingdom.

THE LORDS, REGULATORS, AND ROSLIN

THE YEAR BEFORE Schatten published research that cast doubt on the viability of human cloning, I announced that the Roslin would apply with Christopher Shaw from the Institute of Psychiatry, London, for what was then widely expected to be Britain's first license to clone human embryos. I first discussed it in public in 2002, at a conference in Berlin. Although cloning human babies is outlawed in Britain, therapeutic cloning was made legal that year, after both houses of Parliament voted in favor of it by clear majorities. Little did we realize that the Roslin project would face many delays. The first submission to the Human Fertilisation and Embryology Authority (HFEA) was to come from another British group, while the first (apparently) convincing human cloned embryos were to be produced on the other side of the world.

Our first setback arose because we had to wait for the dust to settle from legal challenges to the HFEA by pro-life campaigners. This setback was entirely acceptable, since it was crucial to establish a clear legal path for the research. Another cause of delay was the HFEA's adherence to the recommendations of a Lords select committee that stipulated that embryos not be cloned for research unless there is an exceptional need that cannot be met by surplus embryos. The implication that a specific justification for cloning embryos is required appears at odds with the admission by the Lords that cloned embryos offer a useful research tool. The project slipped back still further when I became bogged down in paperwork, one of the greatest problems facing researchers in Britain today (although I fully accept that some paperwork is needed for such complex issues, with major social implications). In order to obtain permission to carry out this research, it is necessary to submit the proposal and all the associated forms to ethics committees at the place of the research and any hospital that may supply eggs or cells for use in nuclear transfer. These will be assessed for scientific validity and ethical acceptability. Having cleared the local hurdles, one sends the forms to the HFEA, which circulates them to outside referees for comment upon scientific, clinical, and ethical assessment. The applicants get the opportunity to respond to any comments. All of these comments are then submitted to a final licensing committee. The first, informal contact between our group and the HFEA was on 11 October 2001, but it was not until 8 February 2005 that our license was granted. Many people, myself included, contributed to this delay, but it would surely bring despair to patients who hope to benefit from research in this area.

The target of our application was to create entirely new opportunities to study motor neuron disease, an awful degenerative illness—actually a group of related diseases affecting the "motor" (movement) neurons in the brain and spinal cord. In some countries the more common form is known as amyotrophic lateral sclerosis (ALS) or Lou Gehrig's disease, after the famous baseball player who was a victim. These neurons are the nerve cells along which the brain sends instructions, in the form of electrical impulses, to the muscles and extremities over distances of up to a meter. Degeneration of the motor neurons leads to muscle weakness and wasting. This occurs in arms or legs initially, some groups of muscles being affected more than others. Some people may develop weakness and wasting in the muscles supplying the face and throat, causing problems with speech and difficulty chewing and swallowing. Eventually, breathing is affected.

Understanding of the genetics of the disease is lacking, as is the extent to which various nerve cell types are affected. Mutations in one gene, called SOD-1, have been linked with the disease, causing an accumulation of the abnormal SOD-1 protein in cells, as occurs in other diseases such as Parkinson's and Alzheimer's. But this gene is responsible for only 2 percent of all cases. The mutations responsible for another 8 percent of inherited cases have not been found. In 90 percent of the cases, genes also play a role, albeit a weaker one, but these cases are "sporadic," and their development is thought to be influenced by environmental factors.

More than five thousand people are affected in Britain alone, of whom one thousand die every year, usually of respiratory failure. Diagnosis of motor neuron disease can sometimes take longer than a year. No effective treatments exist, though more progress has been made in the past decade than in the preceding century. To boost the prospects for new treatments and methods of diagnosis, we wanted to collaborate with the Institute of Psychiatry in London and use nuclear transfer to create human blastocysts to derive normal and affected stem cell lines.

We intended to hone methods to create motor neurons reliably from them. The idea was to create at least one normal motor neuron cell line, one with the SOD-1 mutation, and others from patients with another inherited form of motor neuron disease to allow comparisons to be made. We sought to pin down exactly the changes that occur, particularly in the earliest stages. We hoped to lay bare the effects on gene use, protein manufacture, and cellular metabolism.

Chris Shaw on the left and Paul De Sousa back right discuss the project to study ALS in the laboratory at the Roslin Institute with Ian Wilmut. (Photograph courtesy of the Roslin Institute.)

We wanted to find the pathway that leads a healthy motor neuron to decay and die.

We envisaged other benefits that would spring from this research: a line of nerve cells that suffer from motor neuron disease could help develop better tests for the disease, so that it could be diagnosed early enough to give any forthcoming treatments a chance to work. They could be used to screen any new drugs under development; up to 100,000 possible drugs could be tested each year at a cost of up to £20,000. This would be a bargain in terms of money and time: tests of one drug in clinical trails on five hundred patients take two to three years at a cost of £10 million. In this way the cells offered an alternative to animal testing, although opponents of any kind of cloning would, of course, find the idea of "embryos for medical testing" as objectionable as "embryos for spare parts."

In the long term, stem cells also offered a new kind of treatment in their own right: rodent experiments suggested that stem cell transplants into the spinal cord could help prevent decline. The implanted stem cells manufacture cell survival factors that help their neighboring cells and in themselves help repair damage. Hans Keirstead and his colleagues at the University of California, Irvine, for example, have shown how they can obtain from human embryonic stem cells almost pure populations of the specialized cell type that insulates nerves (oligodendrocytes). Injections of these cells improves mobility in rats with spinal cord injuries, providing evidence that these cells can help restore motor skills lost from acute spinal cord tissue damage.

THE FIRST CLONED HUMAN STEM CELLS?

As we were still wrangling with the regulatory authorities over permission to start our efforts to clone human blastocysts, we were overtaken by other groups. The first supposedly convincing evidence that a stem cell line had been produced from a cloned human embryo was produced by a South Korean group. Before their work was published, in 2004, claims to have created a human clone had been made several times already. This time, however, details of how to do it were published in one of the world's most respected science journals. They were reported by a reputable team in Korea led by Woo-Suk Hwang, not, as was occasionally the case, by a cult, a self-publicist, or a maverick scientist. And the human clones that the team produced were much more advanced than those from other earlier attempts by reputable scientists, having been grown to the blastocyst stage.

Other teams had tried to clone human embryos and found that they failed to develop properly, only managing a few cell divisions. The American company Advanced Cell Technology, ACT, claimed, in *E-Biomed: The Journal of Regenerative Medicine*, to have made at least one six-cell embryo by means of the Dolly technique. But, at five days old, this embryo consisted of far fewer than the hundred or so cells one would expect and was probably dead. Nor was there even evidence that the DNA in the clone had come from the donor cell.

The lack of truly novel science in the paper prompted a walkout on the editorial board of the journal by Robin Lovell-Badge, a highly respected stem cell scientist from the National Institute of Medical Research in Mill Hill, London, and two already familiar figures, Davor Solter and John Gearhart. It was not that the work was wrong or fraudulent but that it did not represent an advance. "There was no scientific reason for publishing this. I could not believe that the paper had been refereed properly," Lovell-Badge said at the time. Gearhart complained that the announcement was aimed at investors and the public but not at scientists. "I doubt that these scientists have any problem with the data in the paper," responded Michael West, president and CEO of Advanced Cell Technology. "As I understand it, they disagree that such preliminary results should have been published. On that issue we simply disagree. Our belief is that the data on this research needed and continues to need absolute transparency with the public."

In 1998 a Korean group claimed that it had cloned a human

embryo by nuclear transfer, but its experiment was terminated at the four-cell stage, before the DNA is put to work in the developing embryo, and so they had no evidence of successful reprogramming by nuclear transfer. Then there was the curious case of the Chinese team that claimed to have succeeded in putting human DNA into rabbit eggs, though again some scientists were skeptical. Similar independent studies in Japan have shown development to the blastocyst stage after transfer of monkey nuclei into rabbit eggs.

Despite Gerald Schatten's fear that humans were particularly hard to clone, the obstacles were overcome by Woo-Suk Hwang and colleagues in South Korea. Details of his experiments were published in the journal *Science*, and Hwang, with his colleague Shin-Yong Moon, presented the findings at a feverish press conference held in February 2004 at the annual meeting of the American Association for the Advancement of Science, in Seattle. Hwang read to the journalists from a written note that declared, "Our goal is not to clone humans." Coauthor Jose Cibelli of Michigan State University, who had made intellectual contributions to the manuscript along with a genetic analysis of nonhuman primate cells (all human experiments were done in Korea), remarked at the time, "When I first saw their manuscript, I almost passed out. This was it. This paper proves that you can take a cell committed to being one type of tissue and make it go back in development. This is the first time it was done in humans."

Korea had a powerful stem cell research effort, with an estimated four hundred scientists working in the field. While Hwang was based in Building no. 85 at Seoul National University, Moon, a gynecologist, worked in the Seoul National University Hospital, on the other side of the city. The two labs received most of their funding from the government, a relatively modest $10 million a year.

Hwang grew up in Chungcheong Province, three hours from Seoul, amid the poverty of the Korean War and its aftermath, studied veterinary science, and made his reputation investigating the reproductive biology of cattle, moving to human material after the American stem cell breakthrough in 1998 (still working on pigs for xenotransplantation and on BSE-resistant cattle). Hwang's team seemed to have produced more human embryonic stem cells—most of them not cloned—than any rival team elsewhere in the world. A scientific superstar in his country, Hwang was awarded a decade of free travel for himself and his wife by the nation's flagship airline. Domestic media lionized him as a scientist with "God's hand."

For their first human cloning study, Hwang's team collected 242 eggs from sixteen unpaid women volunteers (or so it was claimed). In a laboratory where the patient grind of cell surgery was combined with automobile assembly line methods, scientists dressed in blue jumpsuits, masks, and bonnets worked on micromanipulators to transfer into those eggs the DNA from the women's own cumulus ("cloud") cells, which surround the ovaries and once helped nurture the egg's development. The dedicated team worked almost every day of the year. When human eggs were not available, it practiced with cattle eggs from slaughterhouses. The team used a chemical to trigger a surge of calcium in the egg, mimicking what happens on fertilization by sperm, making the cloned embryo divide and start development.

By tweaking the time that elapsed between the transfer of the nucleus and the activation of transplanted DNA, Hwang was able to optimize its results: a two-hour delay seemed to work best, so that 20 percent of all reconstructed eggs formed blastocysts. The group had thirty successes and attributed them to stringent timing and to the use of extremely fresh donor eggs—offering a huge advantage over embryos that are surplus to, or rejected from, IVF—and a special method for gently extruding rather than sucking DNA from eggs.

The human blastocysts were cultured for a few days until they consisted of 100 cells, of which 25 were the "inner cell mass," the embryo proper. From 20 suitable inner cell masses, the Korean team managed to obtain only a single human embryonic stem cell line— and thus identical tissue for one woman donor. To demonstrate the flexibility of these much sought-after cells, the team found that this line could turn into three of the main tissue types—ectoderm, meso- derm, and endoderm (*ecto-*, *meso-*, and *endo-* come from the Greek for outer, middle, and inner, while *derm* refers to skin)—that appear at the beginning of development, when the embryo is called a gastrula and organized into three layers.

When transplanted into mice, the stem cells differentiated into still more specific cell types, offering further proof that they were stem cells. The team verified that the embryonic stem cell line was genetically identical to the cell used for cloning by DNA fingerprinting; to make sure it was not the product of parthenogenesis, it could check the use of imprinted genes. But they only succeeded with cumulus cells and eggs from the same woman.

There were other qualms about the initial Korean work. They had managed to obtain 242 eggs, free, from volunteer donors. This raised the eyebrows of American rivals, who would have had to pay thousands of dollars. "Our ethics board never would have allowed that," said one. "They told us if you don't pay them anything it's exploitation, and if you pay too much it's coercion." Some ethicists said they wanted to know more about how the South Korean team attracted the sixteen volunteers. Hwang had explained, "Some young ladies have a lot of curiosity about reproductive cloning and therapeutic cloning, and after searching the Web site they contacted us." At the time Hwang vehemently denied that women in his team donated the eggs. And he stressed that his colleagues had indeed followed guidelines.

Any lingering doubts about the team's efforts were swept away a year later, when the team refined the technique and showed for the first time that it could be clinically relevant. In a paper published in *Science*, Hwang (and twenty-four coauthors) described how eighteen women donated 185 eggs and how, with cells from donors, they were used to clone embryos and then create lines of embryonic stem cells. Predictably, the American President voiced his concerns about where this work was leading: "I worry about a world in which cloning becomes accepted."

The cells to reprogram the 185 eggs had been obtained from eleven donors, who included males and females ranging in age from two to fifty-six, including individuals who had juvenile diabetes, spinal cord injury, and a genetic immune deficiency called congenital hypogammaglobulinemia. This time there were no quibbles about consent. Nuclear transfer was used to create thirty-one embryos, which were used to create eleven lines of human embryonic stem cells, marking an average of 17 eggs for each stem cell line. Among egg donors under thirty, an average of fewer than 14 eggs were used to generate a stem cell line, a stunning level of efficiency—sixteen times better than before. They also provided evidence that the cell lines matched the patients' cells and did not have a parthenogenetic origin, when unfertilized eggs divide on their own.

Success depended on many factors, said Gerald Schatten, coauthor of the Korean work: human "feeder" cells—not animal cells—were used to grow the embryonic cells; Koreans, experienced with handling tricky steel chopsticks, seem more adept at micromanipulation of eggs and embryos; the eggs were from young donors, of high quality, fresh, and handled more gently; the team avoided the use of antibodies and chemicals to obtain cells from the embryos; and the donor DNA was retrieved with a light touch. Hwang himself considers it also important that at least one of his researchers keeps the precious cells company all day and most of the night, as a way of nurturing respect for them. "You need the heart and the spirit, the human touch," he said.

SCANDAL

HWANG WAS BY NOW so popular that he would be mobbed if he went out to eat at a restaurant. The veterinarian joked that he had become Korea's answer to Elvis. He was indeed a rock star of science, one complete with an online fan club entitled "I love Hwang Woo-Suk." The poor boy from a desolate village who had risen to become a giant of cloning by working seven days a week, fifty-two weeks a year (his motto: "No Saturday. No Sunday. No vacations"), inspired at least ten popular books in Korea.

This source of immense national pride was about to become a national embarrassment. Triggered by a report in the journal *Nature*, there had been concerns about exactly how Hwang had obtained human eggs for his research. In the closing months of 2005, the matter came to a head when Gerald Schatten announced he would suspend his collaboration because of "misrepresentations" about egg collection "and the resultant breach of trust." Days later, in an emotional press conference held in Seoul, Hwang admitted that two female scientists in his lab had given their own eggs for research. Under commonly observed international guidelines, this is prohibited so that vulnerable scientists cannot be put under pressure undergo what is a painful and sometimes risky procedure. Hwang apologised profusely, confessing that he had lied when the claim was first made in *Nature*. "I am very sorry that I have to tell the public words that are too shameful and horrible," he said.

The 242 eggs that Hwang had used to create the first human cloned embryo had come from sixteen women, said by him to be unpaid donors. The Health Ministry confirmed that one of the doctors (later revealed to be coauthor Roh Sung Il of MizMedi Hospital), who had earlier been collecting eggs for Hwang's research, paid thousands of dollars to some women for their eggs, which was not then illegal in Korea but is now. And it said that an investigation found that two junior scientists had given their own eggs for research, confirming that the donations weren't in violation of the then-current ethics guidelines because they were made voluntarily. Hwang still maintained, however, that when two researchers offered their eggs he had turned them away. According to him, in 2003 they went behind his back and donated eggs under false names. He added that he had publicly denied reports that this had happened because one researcher asked for her privacy to be respected.

South Koreans rushed to support their science hero. Protesters picketed the headquarters of South Korean broadcaster MBC after the network aired a TV programme that focused on the egg donations by the junior researchers. There was a jump in the number of South Korean women volunteering to give their own eggs for Hwang's work, with hundreds more coming forward. Women even left bouquets of the national flower, hibiscus, at his laboratory, along with touching notes of support.

The hero retreated to a remote Buddhist temple, then, suffering from exhaustion, was admitted to a hospital. Slowly he recovered. But when a weary Hwang returned to his lab early one Monday morning, after an eighteen-day absence, the scandal was far from over. Schatten had by then asked the journal *Science* to take his name off the cloning study published by Hwang and colleagues in May 2005 in which they described how they had used nuclear transfer to make stem cell lines. "My careful re-evaluations of published figures and tables, along with new problematic information, now cast substantial doubts [on] the paper's accuracy," Schatten said in a letter to *Science*. "Over the weekend, I received allegations from someone involved with the experiments that certain elements of the report may be fabricated."

Internet sites buzzed with allegations about the 2005 paper. One question was whether photographs, described as showing stem cells derived from cloned human embryos, were frauds. Roh Sung Il said that they were not derived from cloning experiments. Another question concerned the veracity of the DNA fingerprints used to show that a stem cell was genetically identical to a person who provided cells for cloning. They were "too clean," said one researcher. Another talked of "unusual similarities." However, in a nationally televised news conference, Hwang said that the data were essentially correct and that he stood by his findings.

I had been through a similar experience with Dolly and I knew what had to be done. I added my signature to those on a letter asking Hwang to submit his work for independent scientific confirmation.

By now the unfolding saga had triggered anger and bewilderment among the public. But there was still a bedrock of support. In mid-December the "I love Hwang Woo-Suk" Web site declared: "No matter what anyone says . . . our love for Professor Hwang Woo-Suk will never diminish."

Even then, I too was still convinced that Hwang's team had indeed succeeded in cloning human embryos. I had visited his laboratory and been impressed by the scale of his effort, which had been backed by more than $36 million from the South Korean government. I sympathized with his dilemma: I myself had to do follow-up work to convince high-profile skeptics in America that Dolly was indeed a clone of an adult cell. I also know only too well how media reporting can spin out of control. Then I found out that in January 2005, Hwang had reported to his government sponsors that six of his stem lines had been lost to fungal contamination, an important fact not mentioned in his report later published in *Science*. I felt uneasy.

An expert team from Seoul National University looked into the disputed paper. On 23 December, the investigation delivered its preliminary conclusions. I was on holiday at the time but the shattering findings were inescapable because they made headlines worldwide: the data for the eleven collections of stem cells Hwang claimed to have made were derived from just two stem cell lines. "Based on these facts, the data in the 2005 *Science* paper cannot be some error from a simple mistake, but can only be seen as a deliberate fabrication," the panel said.

The team found no records of two of the other stem cell lines. Four others had died from contamination, and another three hadn't yet become full stem cell lines. To create fake DNA results purporting to show a match between the clones and patients, the panel claimed that Hwang's team split cells from one patient into two test tubes for the analysis—rather than actually match cloned cells to a patient's original cells.

Also in question was Hwang's remarkable efficiency, by which he managed to use just 185 human eggs to create custom-made embryonic stem cells for the eleven patients. The investigation had, said SNU professor Jung-Hye Roe, "found that there have been a lot more eggs used than were reported," suggesting lower efficiency (the investigation would conclude that between November 2002 and November 2005 the team used 2061 eggs from 129 women—one of the more shocking statistics given the risks these women had taken in being stimulated by powerful hormones (which can be life threatening in 1 or 2 percent of cases) and discomfort they had suffered in the name of science.

Even though these were the preliminary findings of its investigation, the panel's conclusion was damning: "This kind of error is a grave act that damages the foundation of science." A statement issued that day from *Science* bluntly referred to "substantial research misconduct." With junior researchers sobbing behind him, an ashen-faced Hwang apologised but reiterated that his team was capable of the feat.

In light of these astonishing revelations, the panel said it would now also investigate Hwang's other landmark papers—which include another *Science* article in 2004 on the world's first cloned human embryos, and an August 2005 paper in the journal *Nature* on the first-ever cloned dog, called Snuppy (Seoul National University puppy), an Afghan hound. The journals were by then already conducting their own investigations. To establish if Snuppy was a clone, *Nature* had commissioned Elaine Ostrander, at the National Human Genome Research Institute in Bethesda, Maryland, to conduct DNA tests. *Science* had contacted the coauthors of its Hwang papers and asked MizMedi to launch an investigation.

During coffee breaks at laboratories around the planet, scientists gossiped about the scandal. How did fraudulent work end up being published in a prestigious journal? How could one keep fabricated data secret in such a large team? Perhaps Hwang's team cut corners and, confident it was capable of the feat, had assumed it could restore the stem cell lines that had been lost to contamination. Perhaps this affair exposed the weakness of Korea's science culture, where the junior members who operate within a rigid hierarchy are not allowed to countenance failure.

By the end of December, any doubt about the last two of the eleven patient-specific cell lines in the 2005 paper had evaporated: "We cannot find stem cells that have identical DNA fingerprint traces with patients, and Hwang's team does not have scientific data to prove they did harvest patient-specific stem cells," said a spokesman for the panel. All the samples presented for Hwang's paper, and even those Hwang said he had created after submitting the paper, were stem cells extracted from fertilized human eggs at Seoul's MizMedi Hospital. Hwang had claimed his stem cells from cloned embryos had been switched. His accusations were met by counterclaims, and this aspect of the affair is still being investigated as I write.

At the start of January, *Science* announced that the 2005 Hwang paper had been retracted. The final verdict was delivered a few days later by the Seoul National University Investigation Committee: Hwang had indeed personally approved the donation of eggs by one of his graduate students and even accompanied her to MizMedi hospital himself. A form asking consent for voluntary egg donation had been circulated around his laboratory and signed by technicians.

His embattled team's fraudulent experiments reached back even further than previously thought and encompassed the most seminal of Hwang's so-called successes: the first creation of stem cells from cloned human embryos, described in the 2004 *Science* paper. The first report of a stem cell line from a human cloned embryo resulted from one created by parthenogenesis, not nuclear transfer. The committee concluded that the DNA fingerprinting analyses and photographs of cells "have also been fabricated." The fraudulent claims were, the panel said, "scandalous." The South Korean government stripped Hwang of the title "top scientist."

There was, however, some good news in the report. The panel concluded that Hwang's team did have the means to clone human blastocysts. And Snuppy looked genuine, which was impressive given that the dog's reproductive cycle is not as well understood as that of pigs, cattle and sheep. To investigate the world's first cloned dog, the team had obtained samples from the egg donor, blood samples from Snuppy, from Tie, the dog that provided the cells for nuclear transfer, and from the surrogate mother, and engaged three independent test centers for the analyses. To the relief of *Nature*, the team concluded that Snuppy was indeed a clone of Tie and that, "when it comes to animal cloning, with the added consideration for the successful cloning of a dog, Korea seems to be internationally competitive."

Whatever the details of what the group actually did and what it fabricated, at least this depressing saga demonstrated that science is relatively quick at weeding out fraud. Overall, the system of checks and balances in research works. As the panel pointed out in its report, "The young scientists who courageously pointed out the fallacy and precipitated the initiation of this investigation are our hope for the future." But, like many other scientists who had visited Hwang and been impressed by his team, his science, and his dedication, I felt badly let down. Although details of what actually happened are still emerging, the deception meant that embryonic stem cell research was years behind where scientists thought it was—in January 2006 it was all hype and no substance. In terms of creating stem cells from a cloned human embryo, we were back to square one. I have no doubt that the Hwang saga will be seen as a fiasco, one that can only be a setback in a field that has already triggered ambivalence and deep unease.

THERAPEUTIC CLONING FOR DIABETES

WHILE THE KOREANS appeared to be overtaking us, the first application made in the West to clone human embryos, part of an effort at Newcastle University led by Alison Murdoch and Miodrag Stojkovic, had been received by the UK's Human Fertilisation and Embryology Authority (HFEA). By that time British stem cell science was enjoying a "brain gain." Stojkovic had already cloned cows, rabbits, and mice, and came to Newcastle upon Tyne from the University of Munich in 2002, one of several stem cell researchers—including Roger Pedersen and Stephen Minger—who moved to Britain to take advantage of its more favorable regulatory and funding conditions. Stojkovic's team at the Institute of Human Genetics had already succeeded in deriving stem cells from noncloned human embryos and now wanted to produce the UK's first cloned embryo to develop treatments for diabetes.

Although the UK has a liberal regime, such work had to be scrutinized. Five members of an HFEA committee studied the proposal, after inspecting the Newcastle lab and quizzing the team. Others asked questions too. Elliott Cannell, speaking on behalf of the pro-life group Comment on Reproductive Ethics, and himself a diabetic, said, "One of the greatest obstacles for a starter is how does anybody imagine they will ever find sufficient human eggs to cure the 170,000,000 diabetic sufferers worldwide. The HFEA has an absolute legal and moral duty to reject this cloning application." I don't think any scientist engaged in stem cell research would dispute the idea that therapeutic cloning does not look very practical as a mass treatment. But we are confident that it will show other ways to achieve the same ends, as well as create type 1 diabetes cell lines that will accelerate the pace of discoveries that will ease the suffering of diabetics. Even if it helps only a fraction of the 170 million suffers, it will be worth it.

The Newcastle project aimed to use some of the 1,500 to 2,000 eggs left over from fertility treatment annually at the Newcastle Fertility Centre, Centre for Life, with fully informed consent from the donors. As Stojkovic put it, "I completely understand the ethical objections, but we are using eggs that are surplus to IVF treatment, which failed to be fertilized. Instead of being thrown away they have been donated for research. The way I think of it is, why put something in the rubbish bin when it can be used in such a valuable way?"

One purpose of the new project was to improve the overall efficiency of therapeutic cloning. As suggested by Hwang, it soon became apparent that fresh eggs were necessary to make nuclear transfer work. Another aim was to lay bare new details of the disease process by comparing cloned diabetic cells with normal cells. Alison Murdoch, the director of the Newcastle Fertility Centre, voiced the hopes that propelled their research, using the example of a ten-year-old child with diabetes in Newcastle. "The ideal situation is that in ten years time, when he is 20, we could take a skin cell, reprogram it, make safe islet cells [which make insulin] and put them back into that boy so he never needs an insulin injection again. That is why this work is so important."

After the license was eventually granted to the Newcastle team, in August 2004, the first in Europe to allow human embryos to be cloned, the image of therapeutic cloning as a potential cure-all suffered a blow when it emerged that the overall aim of the original application had been questioned by an HFEA committee. Although the HFEA was typically vague about what exactly was causing the bother, exaggerated claims for the possibilities of a treatment—discussed below—could have raised legal issues that the pro-life campaigners could have seized on. As a consequence the Newcastle team dropped the overall aim of its project and focused on nuclear transfer alone. The HFEA eventually, in April 2005, gave permission for the original aim of cloning cells from a type 1 diabetic.

I was pleased that Newcastle had been given the go-ahead for its work—the first in the West—but irritated because there was no difference, in essence, between what the Newcastle team had been allowed to do and what I and Christopher Shaw had asked to do years earlier: the HFEA had shifted from insisting that cloning be used for a specific treatment to allowing it to be used to lay the foundations for research on a serious disease. While I had contributed my fair share to delays, the HFEA's about-face was annoying. We could have started our program of work on motor neuron disease as early as 2002. As it was, we were given permission in February 2005. Even that did not mark the start of our work, only the start of the effort to raise funds for the project. Christopher Shaw and I wanted to collaborate with Hwang's team, but the scandal would cast a long shadow over this project.

The kind of treatment envisaged by the team at Newcastle could have improved my father's life immeasurably if it had been available half a century ago. First, I would take a sample of his cells. They could be from the root of a hair or a swab of his cheek. I would transfer all the genetic material from his cell into a human egg whose own DNA had been removed. After the transfer, the egg with its adult nucleus would be stimulated with a shock of electricity to "reprogram" itself with my father's genes and start growing into an embryo/placenta. Then I would have derived his islet cells from the embryonic stem cells. But there is one complication.

In many cases it is an advantage to be able to grow tissue from a genetically identical embryo. That way there is no risk of rejection. Thus, in my father's case, the islet cells would have been compatible. However, diabetes of the kind that affected my father is caused by destruction of the insulin-producing islets by the person's own immune system (hence it is called an autoimmune disease). And we know from a study published by a team at the University of Minnesota in 1989 that replacement of the islet cells with other genetically identical cells does not work as a treatment. When one of a pair of identical twins developed diabetes, islet tissue was transplanted from the healthy to the diabetic twin. Unfortunately the disease developed again. The immune response attacked the transplanted cells, and it is possible that cells from my father's cloned embryo would also have been attacked.

Initially, at least, it seems there may be no advantage in having immunologically identical cells for treatment of diabetes. But there may be ways around this. One approach would be to genetically alter the stem cells in some way (either by genetic modification or by altering the genes that are used—expressed—by the cells with growth conditions) so that the islet cells they make can be tolerated by the body. By a transplant of modified versions of his own islets cells, my father's body could have manufactured its own insulin, sparing him all the awful complications of his diabetes, from amputations to blindness.

In many other cases the ability to grow cells and tissue that carry no risk of rejection would offer huge advantages. Heart muscle cells derived from embryos could be used to patch a damaged area of heart tissue. This has been proved in experimental animals. There would be a real advantage in using cells that were genetically identical to the patient's so that they did not have to take immunosuppressive drugs. Similarly for any other muscle or skin. This kind of work is one of Dolly's greatest legacies.

There is one problem, however. Some people believe that this is tantamount to killing one person to save another. This complaint has dogged the field of human embryonic stem cell research every step of the way, reflecting the continuing debate over the moral status of the embryo. In response to the launch of the stem cell bank, one pro-life group said that the regulatory agency should do "everything in its power to stop unnecessary destruction of these tiny human lives." Its adherents fear that this kind of work will lead to the commodification and cannibalization of embryos. They maintain that a fertilized egg, and everything that follows from it, is sacred. They oppose everything that I propose to do.

SIX

IS A
BLASTOCYST
A PERSON?

WHAT EXACTLY do we mean when we use the word "person"?
What precisely do we refer to when we talk about a "human being"?
To some people even the hollow ball of cells that forms a few days
after a human egg is fertilized is a "person." This microscopic ball is
a "tiny human being" who deserves the same rights as any other,
including, most fundamental of all, a right to life. The very thought
of producing one by the process of cloning is repellent because it
marks the manufacture of a "new kind of human being." Laboratory
cloning is a "violation of the natural order" and a "repugnant manipulation," one that represents a "trivialization" of a human life.

Scientists like me who want to do this are portrayed by our
opponents as having a "lust for forbidden fruit." We are, in effect,
"playing God" and exceeding our authority. We do not have the right
to create, experiment on, or destroy human embryos. My critics

believe that it is not curiosity and a desire to do good but hubris and "hunger for power" that are driving me and my fellow scientists to drag us all into a grim Frankensteinian world.

Critics are horrified by the thought that a human egg can be fertilized in the laboratory, then grown for a few days for experiments. Or that healthy embryos can be selected for a pregnancy and unhealthy ones discarded. Or that an unfertilized egg can be used for cloning. Some even disapprove of using IVF to overcome infertility. This kind of vociferous opposition to human embryo research has already held back efforts to create new treatments based on stem cells. In the United States, under President George W. Bush, federal funding was withdrawn from studies with human embryos because of opposition from the Christian Right, though a vast amount of private and state-funded work would continue. In some European countries, such as Germany, France, and Ireland, the creation of new lines of human embryonic stem cells was banned, creating problems for the European ideal of free exchange of ideas, equipment, and people across borders. As I write this, scientists are agonizing over whether a German researcher who comes to work on human embryonic stem cells with me or other British teams could face legal action when she returns home. John Harris of the University of Manchester is one of a team of ethicists who are struggling to coordinate the ethical stance across Europe with the "Eurostem" project.

At the heart of the furor lies a long and complicated dispute about the early human embryo. This is an argument about the status of the human blastocyst, an embryo that has yet to implant in the uterus. The draft European ethical guidelines drawn up by Harris and his colleagues recognize this by, in effect, agreeing that the status of the blastocyst is the one thing that everyone can disagree on.

I am no moral philosopher but have also had to think hard about this issue, as I am sure most people have. But the questions that trouble the public are not always the same as those that bother the specialists. The latter, be they scientists or moral philosophers, have jargon, terms, and language of their own, making debate difficult. They may feel that certain issues are already settled. But that does not stop the same questions from being raised in the public debate. I want to focus on the concerns of the layperson, which have been voiced to me again and again in conferences and meetings.

The arguments over the critical milestones of a human life have been driven in great part by discoveries about events in the womb. During the first trimester of pregnancy, the mother sees her periods end and her breasts expand, she feels nauseous, and so on. But in the days before medical diagnostics the first obvious sign of the unborn child came with quickening—the first detectable movements of the fetus between seventeen and twenty-two weeks, described by some mothers as a fluttering feeling. Many societies past and present have regarded that stage as being the point at which the conceptus deserved full protection.

Over the past century or two, medical science has given many important new insights into the miracle of development. The German-Estonian Karl Ernst von Baer (1792–1876), who had previously discovered the ovum (egg), showed that the union with the sperm produced a marked change in its appearance. This provided a point of sharp discontinuity, a line in the shifting moral sands, and it was easier to make a distinction that marked the dawn of a life than at any later time in the development of the fetus.

Birth remains much more important to society than conception, however. This is the time of a baby's first breath, her first look at the world, and her first cry. This is the time that the child is given a name and is accepted as a member of the family. This is the event celebrated every year by birthdays. Yet, even here, science provides new information that can trigger debate.

New ways of taking pictures of the fetus in the womb, with sound waves, have stirred deep feelings in the hearts of many parents of the kind that their great-grandparents would have felt only at witnessing the moment of birth. Ultrasound scans have shown that a fetus under the age of abortion appears to be able to do some things that babies do. Meanwhile, medical advances have continued to lower the age at which premature babies can be helped to survive, so that in one part of a hospital doctors can struggle to save a twenty-two-week-old fetus and in another they can terminate it. In Britain these advances in understanding have quite rightly revived discussion of the morality of abortion at later stages of pregnancy.

Reproductive techniques can be equally contentious. Only the older ones among us will remember the furor over artificial insemination that took place fifty years ago. It arose first in the context of cattle breeding and became more relevant and heated when artificial insemination was then used to overcome human infertility. There was also intense debate about *in vitro* fertilization after the birth of Louise Brown in 1978, debate that has now been forgotten. IVF was viewed as crackpot science and its pioneers as outsiders in the medical world who interfered with an intimate process best left to nature. Like me, they were seen as taking the first step down a "slippery slope" leading to eugenics, deformed babies, and doctors playing God. It is reassuring to see how quickly society can develop an informed view so that a subject ceases to be contentious.

The public debate about the significance of events at the beginning of life is reminiscent of earlier discussion about the end of life. When the first heart was transplanted from one person to another, there was much discussion about the circumstances in which it is acceptable to remove an organ. Much of this debate focused on the effects of brain injury—"ideal" donors having a healthy body (and thus healthy organs) but a dead mind. The problem was how to tell when that hard-to-define property of consciousness had ebbed away.

Again it was new scientific observations that helped people to form a view. Most would accept that we should use reactions to stimulation, measurements of flickers of electrical activity in the brain, and scanners such as functional magnetic resonance imaging (which, as the name suggests, studies brain function) to determine the moment of brain death. And so on. The earliest stages of human life require a similar analysis. What are the characteristics that we consider most profoundly human? When do they emerge?

My proposal to carry out nuclear transfer with human eggs and cells has raised many questions. Various different positions have to be distinguished from one another and be reconciled with judgments on abortion, which is more focused on the status of the fetus, a much later stage of development. How do people form their views on these issues? Do they follow guidance of religious teachings, or do they draw on a different ethical framework? Are these religious or moral issues? I know there are many religious views, passionately held, but for me the debate is a moral one.

We live in a secular society, and the question of research on embryos did not arise before the late twentieth century, so the status of the embryo in religious traditions can only be deduced indirectly from the Bible, Koran, and other religious texts. Over the years religious convictions have evolved and changed, often in the light of scientific evidence. A layperson might gain the impression that the Catholic Church has held a consistent line over centuries on the central subject of human life, but in fact it was only as recently as 1869 that Pope Pius IX determined that human life at conception was sacred, although the events during fertilization would not be well understood for many decades to come.

Even today, however, some claims made by religious groups are clouded by misconceptions about the various stages of human development. That is not to say, however, that I am hostile to people who hold them. If I were, I would not have married my wife, who is now a church elder. Nor do I dismiss religious views. Indeed, the Christian tradition, for much of its history, made a distinction between the moral status of the unformed embryo and the formed embryo, and I feel this distinction still has great validity. Since we know that a morally important person may eventually emerge (a substantial proportion of early embryos—some estimates put it as high as 50 percent—are naturally lost before term), I think it is appropriate to accord a gradually increasing moral status to the embryo or fetus, tempered by the recognition that if there is a "threshold of personhood" there will always be argument about where it lies.

In considering the proposals for my embryo research, we must ask various questions: Are all stages of human development equal? Or does moral status develop from the moment of conception through the embryo and fetal stage to birth of a child? Can the early human embryo suffer? Is the blastocyst conscious? Can it be called a person? A theologian and philosopher—Michael Banner of Edinburgh University—expresses the essence of the debate as follows: "Is an embryo a thou or a that?" If we accept that an early embryo is a "that" and lacks feeling and consciousness, the issue is then whether we owe the early embryo respect of such a kind that we ought not to produce and destroy it in the name of science.

I want you to take stock of what you think about the moral status of the early embryo. Imagine how you would behave in the following scenario. You are visiting an IVF laboratory with a young child. A laboratory assistant brings in a petri dish, carrying it with great care so as not to spill the contents. The child is fascinated to see twelve early embryos in the dish, each consisting of eight cells, under the gaze of a microscope. Two embryos are to be transferred to a patient, and the other ten are to be frozen and stored in a Dewar flask for possible later use. Suddenly, before anything can happen, the fire alarm goes off. Smoke is billowing in from the corridor, and you have moments to squeeze through a small window before the lab fills with fumes. What would you do? Carry the child or the culture dish containing the twelve embryos?

Some believe that the early embryo must be regarded as a human being in the fullest sense from the moment of fertilization, and accorded the same respect as a baby. Thus the destruction of the embryo should be banned. And, presumably, those who hold this view may be tempted to save the dish, for it contains twelve lives. Others regard the early embryo as little more than a collection of undifferentiated cells, deserving no more attention than any other isolated human cell or tissues. And there is the middle ground where the embryo deserves a degree of respect because of the potential to develop into a human being in certain circumstances. If this line is accepted, then we raise another problem: What does it mean to show an embryo respect? Perhaps we can shed some light on these issues by first considering what makes humans special.

WHY ARE
HUMANS SPECIAL?

SOME OBJECTIONS to the use of human embryos rest on the notion that there is something "special" when it comes to human life. I agree that we are special but feel that there is more to this than the basic attribute of being alive. From the perspective of a one-cell creature, to say that a human blastocyst is merely living is to do it a great injustice. Like a bacterium, each cell in a blastocyst reproduces and "eats." But it does something altogether more fascinating, something that deserves profound respect. When its cells split, they don't spawn new organisms but stay linked with the old ones to form a complex, interacting community. One of the greatest mysteries of all is how parental genes in a fertilized egg guide it within the womb to develop into this complex and finely integrated community that is you or me, which consists of two hundred or so different tribes of cells.

We know the process in rough outline. First the fertilized egg splits to become 2 cells. Then they both split once again to become 4. Then 8, then 16, and so on. Repeat this process of cleaving 2 by 2 by 2 around 47 times overall, and you end up with an intertwined and interacting supersociety of 10,000,000,000,000,000 cells. The actual number in an adult person is significantly fewer than this because nature kills off many of these cells by a process called apoptosis, making it hard to estimate the final tally. In this way the body removes webbing between fingers, knits together the palate of the mouth, carves out toes, shapes organs, removes defective cells, and so on. The result is a complicated, cooperating three-dimensional whole formed of (fewer than) 10,000 trillion cells. This cellular community is what we call a person. But it has to be stressed that, although this person began as a blastocyst, a person does not count as a blastocyst any more than a blastocyst counts as a person.

At the level of the cell, are we any more special than other creatures? Certainly more so than bacteria, which have a different design (in technical terms, they lack a cell nucleus). But comparisons of our genetic code with those of other multicellular organisms reveal that many of the genes we carry are dedicated to making the vast collection of 10,000 trillion cells in all these different tribes work together as a coherent whole. The more important the role of genes in running the basic biochemical machinery of life, the more likely we are to share them with other creatures. That is why we share about half our genes with a banana, for example. As a result, at the genetic level we have a surprising amount in common with yeast, apes, and pretty much any creature you can think of.

At this fundamental chemical level, you, I, and the rest of *Homo sapiens* are different but not particularly special given what we know about human evolution and ancestry, let alone the astonishing similarities between our species and others. A reductionist alien scientist gazing only at the workings of human, yeast, and banana cells would think the building blocks of a human are more similar to than different from the building blocks of apes, mice, and a host of other creatures.

However, when it comes to the phenotype of a creature—the way it looks and what it can do—*Homo sapiens* is distinct. Now we have moved from the properties of a cell to the emergent properties of trillions of cells. We may not be able to outrun a cheetah, fly like a bird, or bask in superheated waters near hydrothermal vents, but we are quite distinct and special in terms of what we can do. And there is one key respect in which I do feel humans are special, not in the sense of an attribute that they alone possess but in the sense that they show an extreme manifestation.

I suspect that all animals, from the lowliest insects to the most sophisticated apes, experience varying degrees of consciousness and self-awareness. Precisely where the line should be drawn between a creature that reacts and one that can really think is a matter of intense debate, but there is no argument at all when it comes to the view that human consciousness is a special property of the cosmic numbers of nerve cells packed between our ears. From their activity emerges the property of self-consciousness, the ability to reflect and to reason. This is the crucial point to which I will return in discussions about whether the embryo should have the same rights as an adult.

THE BIRTH OF
GENETIC IDENTITY

IN OUR QUEST to weigh the moral status of the early embryo, let's start at the very beginning of when an individual is created. You might expect that when egg fuses with sperm, this is the very instant that a new individual is born. By this I mean an individual with a distinct genetic identity, compared with that of the egg and sperm. Yet studies reveal that the significance of the point at which egg docks with sperm is much less clear-cut. The genetic cargoes of egg and sperm initially lie dormant and separate, cloistered in different structures called pronuclei. These are residual egg and sperm, if you like, and the DNA of the new individual has not yet assumed control of development. In the early days of IVF, in the 1960s, I can remember a debate being triggered by a photograph published by one of the pioneers of IVF, Bob Edwards. The image clearly showed the two pronuclei, and other scientists present began to argue about whether it really counted as an embryo, whether fertilization had occurred. Even in that scientific context there was debate about what exactly we mean by describing something as "an embryo."

Only when the pronuclei have come together and the fertilized human egg has divided three times do the genes of the embryo take control of development. Only then can we talk about the genetic recipe that will define a person. Only at that time are the instructions in that recipe acted on; before that point, instructions from the mother and father—in the form of RNA and protein—help to direct the first stages of life. Only after this stage does an individual's DNA take over. One could argue that only at this very moment is an individual present in the most basic sense, that of a distinct genetic recipe.

The blastocyst that sits silently at the heart of the stem cell debate is a far cry from the popular images of the embryo, where it is depicted as a little fetus with limbs and organs such as heart, brain, and liver. In reality, this stage forms after around seven weeks of development, when all of the individual organs become recognizable. By contrast, the blastocyst stage of life that forms after a week or so consists of a ball of cells stuck together like luminous tapioca, or frog spawn. Just visible to the naked eye, but smaller than a grain of sand or crystal of salt. When I look at a sheep blastocyst under the microscope, as I have done a thousand times, I do not think of it as a sheep. Nor would I think of babies when I gaze at a human blastocyst, though I accept that some couples undergoing IVF treatment do become attached to their embryos.

Each human blastocyst consists of a hundred or more cells. The much sought-after embryonic stem cells are among the twenty-five or so in the "inner cell mass," a clutch of small cells within the hollow sphere that are destined to become the fetus and some of the placenta. Unlike adult stem cells, they have the ability to turn into any part of an embryo and thus make any cell type in the body.

Embryonic stem cells, like people, respond to their culture, surroundings, and upbringing. But the conditions experienced by embryonic stem cells in a laboratory can never fully mirror those experienced by cells in an embryo developing within the womb. Although biologists have been studying and using mouse embryo stem cells for twenty years now, we do not understand the details of why these cells derived from the inner cells of a blastocyst are such extraordinarily versatile stem cells or how they turn into any other type of cell.

Once separated from their sister cells, these embryonic stem cells lack the crucial cues and signals to develop into a fetus. When grown in a suspended culture of proteins and hormones, for example, they form an embryoid body, a spherical structure consisting of a clump of cells that differentiates in a disordered way, so it does not count as an embryo and cannot develop into a fetus. An embryo needs the environment of the womb if it is to develop at all. In a Petri dish it has no potential to do this.

An early embryo is unlike you or me or any other individual person in other respects. One blastocyst can split into two, as happens naturally to produce twins. Conversely, if there are two embryos, they can combine into one early in pregnancy, as happens naturally in the womb, albeit rarely, to create a chimera. Or a blastocyst can, in theory at least, split into two and recombine again. Did "life" in such a case begin as an individual, become two individuals, and then turn into a singleton again?

There are other peculiarities. What many do not fully appreciate is that three-quarters of the cells in the blastocyst, around seventy-five, are destined to make the first contact with the lining of the womb and contribute to the placenta. As highlighted by the people who use the placenta to fertilize plants, consume it as a postpartum snack (browse the Internet and you will find plenty of recipes), and use it as a medicine, society does not accord the progeny of these cells the same respect as the embryo proper.

None of the above points mean, of course, that an embryo is not an individual. Nor do they suggest than an embryo does not deserve respect. I merely point out the messy details to show that when we do get the full picture of the first stages in a human life, the identity of an early embryo does not map neatly onto the identity of a person.

A BLASTOCYST IS NOT A PERSON

ALTHOUGH A HUMAN embryo has only just begun development, it is deserving of our respect because it has the potential to become a person, in the right circumstances. However, because a blastocyst has this potential does not, of course, mean that the blastocyst has the same authority as a person, just as a young girl who wants to study medicine does not have the same rights as a qualified doctor. As John Harris argued in *The Value of Life*, we are all potentially dead but that does not mean we should be treated as if we are dead.

Nor does a fertilized egg created by scientists in itself have any potential to become a person without human intervention. The early embryo sits not in a womb but in a petri dish and is difficult to grow in laboratory conditions for more than five or six days. Until it attaches to the wall of the uterus, where it starts a highly complex chemical dialogue with the mother, there is no pregnancy. A blastocyst sitting in a Petri dish does not go through further stages of embryonic development. It has the potential to grow into a fetus or a baby, but only with human assistance.

The central reason that I don't regard a blastocyst as a person is that it has no mental life. In an adult person there are around 100,000 million synapses, the connections between nerve cells. Each cell has around 1,000 connections with other cells, so there are about 100 million million connections that store information by a process called plasticity (when one nerve cell "talks" to another, it can modify how efficiently communication occurs, and this is believed to be critical for brain function). A working human brain is thought to make one million connections, on average, every second. The critical issue is when this capacity to think first appears.

In effect, we are groping for the idea of what some call "brain birth," the mirror image of "brain death." When this quality emerges is the focus of much debate. This was highlighted by an expert panel that discussed proposals to grow human neural stem cells in the brains of developing nonhuman primates to provide better models of Parkinson's and other neurological diseases. The panel, which included Davor Solter and John Gearhart, concluded that the implants could affect cognition such that the resulting animal would have a moral status closer to that of humans. Importantly, however, there was no consensus on the moral significance of changes in abilities even if we could detect them. John Harris would argue that a "person" is *a creature capable of valuing its own existence.* This, he says, makes plausible an explanation of the nature of the wrong done to such a being when it is deprived of existence. Over a lifetime an individual will gradually move from being a potential person or a preperson into an actual person when she becomes capable of valuing her own existence. "And if, eventually, she permanently loses this capacity prior to death, she will have ceased to be a person."

This is all very logical and coherent, but I suspect most people would regard an elderly relative who had lost his mind to Alzheimer's as a person. For many the same would go for individuals who are "brain-dead," anencephalic infants, or individuals in a persistent vegetative state. And by the same token I would have an equally conservative view of personhood at the start of human life. Where, precisely, I cannot say. Except that the clutch of one or two hundred cells that makes up a blastocyst is *not* a person.

No feelings or sentience will stir within the blastocyst without a functioning nervous system. Although stem cells are removed from an embryo around six or seven days after fertilization, it is only at around fourteen days, when it consists of several thousand cells and is still no bigger than the point at the end of this sentence, that an embryo develops the "primitive streak," the first glimmer of the machinery of thought. This is the precursor to the spinal cord and backbone and central nervous system that will develop later. It is also beyond the point where twinning can occur.

For me this is an important milestone in development. Now, at last, we have reached a point where the embryo is committed to creating a single human being. Crucially, that point can be defined in an objective way that, in the words of the great theological thinker Dietrich Bonhoeffer, allows me to take up the "sharp swords" of simplicity and wisdom.

This is why I feel I can justify the use of blastocysts in my research. In turn, this raises other questions familiar to anyone who does research on living creatures. Is a blastocyst aware? Can a blastocyst feel pain? Until nerve connections form between two crucial areas of the developing brain, the cortex and the thalamus, sensations of pain cannot be experienced. This formation takes place much later in a pregnancy, around the twenty-sixth week. There is no pain, no suffering, and thus no cruelty to the blastocyst is possible.

RESPECT THE EMBRYO

IN THE LIGHT OF the above discussion, I understand and back the current UK regulations that embryos more than fourteen days old not be used in research. The appearance of the primitive streak marks a very conservative starting point for a person, one that gives the benefit of the doubt to those who feel unease about embryo research. I believe that the use of blastocysts before this can and should be justified in carefully regulated research and in certain circumstances. That is not to say, however, that an embryo younger than two weeks is not deserving of some respect.

Although the respect shown an early embryo falls far short of total protection, British regulations do reveal this respect in several ways that I fully support: early embryos should be used in research only if there is no alternative and only with consent, and meticulous record keeping must ensure that every embryo can be accounted for. One would expect that the research on an embryo must be of the highest quality and justifiable. As part of the respect I accord the embryo, I share with many others the view that we should do our best to use and investigate alternatives, such as adult stem cells or parthenotes, formed by persuading an unfertilized egg to divide.

Despite the deliberations on this issue by the Roslin group, politicians, and many others, I know I will be criticized. At one extreme are those who maintain that my research on human blastocysts is tantamount to killing people. At the other are those who find the above discussion of regard for the early embryo woolly-headed. Regarding my respect for the early embryo, John Harris has remarked, "Is that like saying that I respect the pig in the bacon sandwich and am doing my best to find something I like to eat as much as bacon sandwiches?"

We should be realistic about whether it will ever be possible to reduce complex ethical and moral issues to pure logic, however. The work of the Kurt Gödel and others showed that it is not even possible to deduce all mathematics from the axioms of logic, leading one mathematician to joke that, if we define a religion as a system of thought that contains unprovable statements, and thus an element of faith, "Gödel has taught us that not only is mathematics a religion but it is the only religion able to prove itself to be one." Even scientists who reject religion must have some articles of faith.

DOUBTS, BUT NOT ABOUT THE BLASTOCYST

DESPITE THE best efforts of philosophers, and despite my little contribution to the debate, science as it stands cannot produce a definitive statement on the moral status of the early embryo. One huge hurdle is the absence of general agreement on what we mean by life. The eminent Austrian physicist Erwin Schrödinger suggested in his book *What Is Life?* that a fundamental property was the tendency of life to produce increased order and seemingly unlikely arrangements of things, whereas the second law of thermodynamics suggested that, left to themselves, things tend to end up in the most probable state, a disordered mess. Another proposal is that life is intricately associated with evolution: life is a self-sustained chemical system capable of undergoing Darwinian evolution. Biologists can list a whole set of features possessed by nearly all living things. As well as the ability to reproduce, these include the existence of genetic information, complexity, organization, and so on.

But exceptions can always be found. For example, the ability to reproduce is not possessed by every object we might expect to call "living": sterile men, postmenopausal women, mules, and viruses are all incapable of self-reproduction. A sneeze demonstrates how a common cold virus can persuade cells in the human respiratory tract to reproduce and spread it, even though it is little more than a stretch of genetic programming wrapped in protein. Nonliving things also show some "vital" signs. Crystals, for instance, are capable of self-reproduction during growth. And so forth.

The same confusion surrounds the critical issue of consciousness. All can agree that the human brain is unmatched in its ability to think, to communicate, and to reason. Most striking of all, it has a unique awareness of its identity and of its place in space and time. But consciousness is notoriously difficult to define or locate. Many cultures and religious traditions place the seat of consciousness in a soul that is separate from the body. Conversely, many scientists—I among them—and philosophers consider consciousness to be intimately linked to the neural functioning of the brain. If we could define the moment that a developing embryo became conscious, and quantify this consciousness, that would lay down a crucial marker for the debate. I remain doubtful that this will ever be possible.

Some would argue that I cannot therefore prove, in the strongest sense, that a blastocyst is not a person. Thus the early embryo should be given the benefit of the doubt. Even though probably not persons, they should be treated as such and accorded the full rights of a baby. I would accept that this infinitesimal doubt about the status of the early embryo would indeed be enough to halt research. Once again, however, I have no doubt that a blastocyst does not possess consciousness. Although the arguments in support of embryo experimentation fall short of the rigor of a mathematical proof, there are also morally weighty reasons to continue. I respect the early embryo and have duties to it, but I passionately believe these are overridden by the duty we have to people.

THE RIGHTS OF A BLASTOCYST'S PARENTS

ONE SOCIAL CHANGE in my lifetime that I applaud is the considerable lengths to which we now go to provide equality of opportunity. However, this is very different from saying that every embryo deserves the chance of life, as some argue. Or of suggesting that potential parents should not exploit modern medicine to prevent the birth of children with inherited disease.

Since the embryo is unable to make judgments, it depends upon the "parents" to act on its behalf, as indeed they would if it continued to develop into a child. The debate about reproductive technology often neglects to note that embryo research can continue only with the cooperation of the people who give their approval for their embryos to be used. They have rights, too. Indeed, even if reproductive cloning were allowed today in our democratic society (and I hope it never is), and even if all the scientists wanting to do it were the Nazi eugenicists that the critics fear, it could happen only with the approval of two or three people, the individual who is to be cloned, the woman who donates her eggs, and the woman (perhaps the egg donor) who is to become pregnant. Feminist groups claim that women's bodies are being exploited in an unacceptable manner in these new reproductive technologies. The point is, however, that people do have the right to chose, to say no if they do not want their embryos used this way.

Some parents confront these decisions already; as we learn more about human genetic diseases, more of them will. In many cases the embryo will have a genetic makeup that makes it vulnerable to an inherited disease, or even condemns it to a short life. Even now, IVF embryos can be screened when parents know that an awful disease stalks their family, so that only healthy embryos are implanted and the rest discarded. So-called preimplantation genetic diagnosis (PGD) can deal with disorders with specific gene culprits—including cystic fibrosis, sickle-cell anemia, muscular dystrophy, Tay-Sachs disease, and fragile X syndrome. Newer techniques also have enabled PGD to detect chromosomal aberrations that cause Down syndrome and certain blood cell disorders and so on. The possibilities are expanding, and the use of these tests is a matter for the parents to decide: the impact of a test for a mutated gene, whether positive or negative, can have consequences not just for the embryo but for the family at large. And some of these are not always obvious. Think, for example, of the middle-aged woman who finds out from her identical twin sister, who is undergoing PGD, that she does not carry a gene for the devastating degenerative disorder that killed their mother: she may well have mixed feelings if she has spent her life avoiding relationships with men and the pressure to have children.

What of surplus embryos that are produced during fertility treatment that are not known to be vulnerable to inherited disease? Sometimes a couple will be fortunate that so many healthy embryos are produced that they have some to spare even after they have had children. These are likely to be frozen and stored so that the decision as to what to do with them can be delayed until years after their production. The couple may choose to donate them to other infertile couples who are not so fortunate, but this requires their accepting that someone else will bring up those children. Others may choose to donate the embryo for research, as has been done to create embryonic stem cell lines.

Finally, the parents may decide to destroy the embryos. If the couple cannot contemplate the idea either of a child of theirs being born in a different family or of embryos' being used in research, this is the only option available to them. In Britain the law requires that frozen embryos be stored only for five years before they are thawed and either used or destroyed. The couple would have been made aware of this deadline before fertility treatment began.

Many people take a pragmatic view. They may not support the production of embryos specifically for research or therapy, but accept that embryos that already exist should be used rather than destroyed. As a result of the astonishing rise of IVF over the past two decades, many blastocysts are now left over from fertility treatment. Of the 763,509 embryos created by IVF between 1991 and 1998 in Britain, for example, 184,000 were stored, 48,000 were used in research, and 238,000 were destroyed. Pragmatists argue that it is better to make use of this precious resource—with the informed permission of the couples concerned—rather than throw it away. They argue that this accords the blastocyst greater respect. Indeed, they say, using them in research to develop new treatments and deeper understanding of serious disease is more than morally defensible: it is morally required.

What, then, of the production of embryos for the specific purpose of research or therapy? In the dozen years after the enactment of British legislation introduced in 1990, only 118 embryos were created for experiments, compared with the hundreds of thousands used in IVF. And rightly so, because the specific creation of a blastocyst for research should be a last resort in exceptional situations.

Once again, precise circumstances may vary. A healthy woman may chose to donate her eggs in the knowledge that they will be used not to produce a child but to create cloned embryos with an inherited disease that offer new opportunities to understand the illness, test drugs, and so on. One embryo provides endless opportunities for research since, once a stem cell line has been established, it can be grown in the laboratory for years. Perhaps she will choose to do this because she has seen the suffering of a child, friend, or colleague and understands the potential value of the research. Others know firsthand the suffering that this research could help end. My group is asking people with inherited forms of motor neuron disease to provide cells so that we may use them in nuclear transfer to re-create in the laboratory the cells that were present in their bodies early in life. In this way we hope to gain the first glimmer of the fundamental cause of what is going wrong. The donors recognize that there is only a limited chance that the research will assist them, but they want to help future patients.

Whatever the motivation, it is crucial that donors be well informed. That information must come from an impartial source, someone who does not stand to benefit from the research or therapy. Nor should there be any question of financial reward for donors. The circumstances will often be difficult and charged with emotion. The responsibility of the counselor, typically a nurse, is to describe what will be involved for the donors—the physical discomfort they will suffer and the risk of any long-term effect upon their health. The donors must understand exactly what will happen in the laboratory and what stage of embryo will develop. Those people who find it unacceptable to produce or destroy embryos must not be pressured to change their minds. There must be time for the potential donors to weigh all the information and then even to reconsider. Cell donors have a basic right to say no.

THE YUCK FACTOR

FUNDAMENTALIST OPPONENTS raise many other objections to signal their disgust with embryo manipulation. Some claim that therapeutic cloning and test-tube babies diminish our humanity. Hardly. The extraordinary inventiveness that made nuclear transfer and IVF possible is a defining human characteristic. Many critics charge that embryo research is an affront to human dignity. But this argument seems to founder on what we actually mean by human dignity. Human beings deserve respect, but, then again, so do all living things with the ability to feel and think to some degree. Is it undignified to use one's ingenuity to overcome infertility, to treat disease, and to reduce suffering? I don't think so. In fact, the opposite is true. Surely cloning replacement nerve cells to repair a dementing brain is more dignified than leaving an Alzheimer's patient to lose his memories, his friends, his home, and even control over his bowel functions.

Many critics believe that my research is "unnatural." This complaint raises various questions about what we mean by "natural." When it comes to our home planet, for example, which vintage of Mother Earth is the most "natural" and thus which one should we be faithful to? Earth as it was 100 years ago? Or Earth 1,000 years ago? Or Earth before bacteria had enriched its atmosphere with oxygen? Or perhaps a very new Earth, one subjected to the hell of the great bombardment 3.9 billion years ago, when our newborn planet swept up the garbage and debris out of its orbital neighborhood?

The very idea that some things are natural and others not raises more questions. If humans are natural, are the tools they develop natural too? And if not, does that mean the pandanus leaf tool of the New Caledonian crow is unnatural? And how about the "fishing sticks" of apes and those rocks that Egyptian vultures drop on ostrich eggs?

Many things that are commonplace in nature seem, well, unnatural, when revealed in detail by scientists. When people have a streaming cold, a scrap of genetic material wrapped up in protein (a virus) has invaded their respiratory tract, genetically altering their cells so they become virus factories. To many people the thought of alien genes taking over their noses may sound fantastic. But it happens every day. They may also be perturbed to discover that their cells are powered by microbial technology—structures called mitochondria, which were once independent organisms but then, some two billion years ago, set up home in the cells of our ancestors, trading energy for a comfortable and dependable home. They also rely on pounds of microbes to digest their food. Indeed, some of these microbes interact with intestinal cells, called Paneth cells, to promote the development of blood vessels in the intestinal lining following birth. It sounds incredible—like a B movie science-fiction plot or a tabloid headline—but, from princes to paupers, we are all bug powered.

Many people still hanker after a more natural world. I would sympathize with them in the case of a hunter-gatherer who has no dealings with computers, cars, antibiotics, disinfectants, and all the other "unnatural" trappings of modern life that make us live longer and more comfortably than ever before. But in the developed world the assumption that the way nature intended it is best is a fallacy. In a state of nature—or an earlier time in our history that more closely approximates it than that of mankind in late industrial society—there would be many more deaths in childbirth, many more people killed by infection, and much more suffering. No one suggests a return to such a state on the grounds that it is more natural. No one mourns the passing of smallpox, either. No one looks back to the good old days before pain relief. When air-conditioning our homes, driving to work, and taking acetaminophen when we have a headache, we conveniently ignore this fallacy in our daily lives, but the very mention of cloning often triggers passionate cries that it should be rejected because "nature knows best."

I could even argue that cloning (though not the process of

nuclear transfer used to create Dolly) is the most natural thing in the world. After all, in the very beginning was the clone. Four billion years ago, replicating chemical reactions stirred on the newborn planet Earth. The seeds of life were sown by these reactions in which molecules reproduced themselves to make "chemical clones." They began to straddle the watershed that divides multiplying molecules, simple replicators, and other types of dead chemistry from animate, breathing biochemistry. Something emerged that we would today classify as living—a self-replicating chemical reaction wrapped within a protective filmy envelope. Within this first organism, billions of dumb and purpose-free molecules engaged in a complex choreography, enabling it to cleave. Then one became two, and the first clone was born.

Echoes of the chemistry that turned within that very first clone can still be found within our own cells. The cleaving when the cells in our bodies divide can be thought of as a kind of cloning. The duplication of the genes in our cells during cell division can also be considered cloning. Each of our body cells is a clone of all the other cells and of the original fertilized egg. Mitochondria, the chemical power packs that drive our cells, clone themselves to help drive on the next generation. We owe our most ancient origins and our continued existence to cloning.

Then, of course, there is the "clone" that arises from a technique called nuclear transfer, such as Dolly. This is a highly inefficient method for making a genetic copy of a creature, a second-rate imitation of the process that bacteria have carried out effortlessly for billions of years.

Of all the cries of "yuck," one of the loudest comes from those who worry that if cloning embryos for therapy is allowed, we will inevitably end up with cloned babies. Insofar as using a six-day-old embryo in research places us on a slippery slope, there is nothing special in that. Our heels are skidding on a steep and slippery slope in our treatment of animals, in our legislation on abortion, and in that which governs sexual behavior. It is the job of democracy to decide just how many inches down the slope we should slide. I cannot deny that research on therapeutic cloning will, inevitably, help hone techniques that can be used in reproductive cloning. But that does not mean that the latter will happen. Reproductive cloning is banned in Britain, while therapeutic cloning is allowed under strict regulation. No matter how slippery that slope, it is easy to draw a well-defined line on it in this particular case. Society can have the good and reject the bad.

POWER, RESPONSIBILITY, AND PROCREATION

THE BASIC FEARS stirred by the opportunities created by scientific research are not new. The city where I live once saw protests at the introduction of anesthetics during childbirth. Professor James Simpson was born in 1811 to the family of a poor Edinburgh baker, the seventh son of eight children. He was both very intelligent and deeply compassionate. Appointed professor of medicine and midwifery at the remarkable age of twenty-nine, he introduced the use of ether and chloroform in obstetrics. But he faced opposition.

In the early days anesthesia could be hazardous. To understandable fears about safety were added religious objections. Pain was thought to be necessary for the normal progress of labor, while churchmen (and some physicians) held that anesthesia, in abolishing the pains representing God's punishment on womankind for tempting Adam's fall, was sinful because it circumvented the chastisement inflicted by a Higher Power. Local clerics, who of course were all men, argued that women were intended to suffer in childbirth.

But some religious leaders were supportive, and Simpson gradually won the argument, especially after Queen Victoria's use of chloroform for the birth of her eighth child, Prince Leopold, in 1853. The local clinic still bears Simpson's name, and it is said that when he died thirty thousand mourners lined the streets of Edinburgh. A plaque in Westminster Abbey reads, "To whose genius and benevolence the world owes the blessings derived from the use of chloroform for the relief of suffering."

Similar debates have raged about the application of knowledge ever since our ancestors learned how to kindle fire. To ban all cloning—both therapeutic and reproductive—is akin to a Stone Age tribe's banning any use of fire because you can burn down your neighbor's hut, even though it also keeps you warm and sterilizes your food. The understanding of how an influenza virus killed up to fifty million people in 1918 can just as easily be abused to make a biowarfare agent as used to create vaccines and drugs to halt a future pandemic. And the biology that will let people clone children, which may have unfortunate consequences for them (if they survive childbirth), will also help us to treat some very nasty diseases.

Every time I hear calls for blanket curbs on genetic technology, I think of the heart patient who died waiting for a transplant as animal rights protestors championed the rights of xenograft pigs, or the mother forced to watch her son suffer because fundamentalists took legal action to prevent her from having what they call a "designer baby" (a technique that relies on embryo selection, not genetic modification, as this pejorative label suggests). Abandoning a particularly "dangerous" technology wholesale can kill, maim, and hurt future generations by preventing that technology from doing any good at all. Society has to weigh the opportunities to help and make sure that it does not miss important new opportunities because of fear of new knowledge. We should expect to change our views and judgments in the light of new discoveries.

Each advance in reproductive science sends shock waves through society. Perhaps the best-known example is IVF, where an egg is fertilized in the laboratory and the resulting embryo implanted in the mother. Indeed, we would not be arguing about the status of the embryo without this research. Apart from bringing hope to millions of childless couples, IVF has provided access to the hidden world of the human egg and the embryo.

If we have learned anything from Louise Brown, the world's first IVF baby, it is that the revulsion that greets cloning is likely to give way to acceptance of some, but not necessarily all, of its uses when the public becomes aware of the benefits, such as the ability to grow replacement cells to repair a damaged body. Many of the major advances in biology have evoked that same knee-jerk outpouring of horror. The first deliberate transfer of genes between microbes and the first heart transplant are among them. Now they are everyday. I would like to think that the same will go for use of nuclear transfer to derive cells or to correct human genetic disease. As I hope to explain later, there is one application of cloning to produce a child that may one day offer real benefits to society by eradicating genetic diseases.

WHY WE
SHOULD NOT
CLONE BABIES

WHEN LESLEY BROWN held her daughter Louise for the first time, she was speechless. The expression on her face was absolutely extraordinary, remarked one of the doctors present at the birth in Oldham and District General Hospital. After about a minute she turned to them and said, "Thank you for my baby." John Brown refused to hold his little daughter at first. He was shaking too much, on what he now calls the most beautiful day of his life. For scientists, too, 25 July 1978 was historic. On that day was born a healthy baby girl who was conceived in a glass Petri dish—the world's first test-tube baby. Since Louise more than two million babies have gazed upon parents who would otherwise have failed to conceive children.

Louise Brown's birth helped launch a revolution in reproductive technologies and medicine that is still changing the world, challenging moral values, and stirring religious debate. As one of the pioneers of IVF, Robert Edwards, said of their impact, "They change an attitude. They change a way of doing things. They change an ethic. They change the way a nation looks at itself." Louise Brown's story remains extraordinary for the insights it gives into the relationship between society and reproductive science, when the despair of childless couples—who constitute between 9 and 15 percent of all married couples—met the ambition and scientific curiosity of doctors.

Her story is highly relevant to the debate over cloning. Before Louise Brown was conceived, human reproduction was regarded as almost sacred. Edwards said it seemed to be a matter "for theologians, rather than gynecologists and scientists." Despite the misgivings, IVF has become widely accepted because alleviating human misery is the driving force, as is the case with therapeutic cloning.

But will that also be the case with reproductive cloning, the "copying" of people? While millions will be disgusted by the idea, there will be a few desperate people for whom cloning will raise hopes, all false. There will be some, like the biologist Richard Dawkins, who have genuine curiosity about cloning themselves ("I find it a personally riveting thought that I could watch a small copy of myself, fifty years younger"). There will be a small group of doctors hoping to exploit these desperate people to make a living and to earn their place in history. Others will talk about the need to be pragmatic and about circumstances in which it would be justified to clone a baby. But, as Bertrand Russell once said, "Pragmatism is like that warm bath that heats up so imperceptibly that you don't know when to scream." Many places will have legislation to ensure we never reach the boiling point of reproductive cloning. However, laws forbid; they do not always prevent.

I do not believe that the moral of the story of IVF is that anything in reproductive science goes, including the cloning of babies. As I argued earlier, in discussing Stephen Levick's work, there are good reasons not to clone. People have already cloned dead pets in the mistaken belief that the resulting genetic twin is a kind of "resurrection" of their lost animal. Everything from the coat color to the behavior of these clones has underlined how this is not true. Because there is no such thing as genetic determinism, a clone could never replace a lost child. Levick has also explained the identity confusion that would arise from a child's knowing she was her mother's identical twin, or perhaps even her grandmother's. Others have warned of discrimination against clones, or "clonism."

Crucially, there was a real point to IVF, which had been driven by efforts to alleviate a common cause of heartbreak. Louise's mother and father, Lesley and John, were the one couple in eight who could not have children. According to the British Fertility Society, nearly 2 percent of all babies born in the UK during 2005 were conceived thanks to assisted reproductive technologies. The technique has given millions of people hope.

Some advocates of reproductive cloning have claimed that infertile men and women could now resort to nuclear transfer to have a child, or that homosexual couples could produce a child from a chimera blended from two cloned embryos. But I find it hard to imagine a scenario where the huge risks of cloning—outlined in detail below—can be justifiable, given the range of alternatives on offer: old-fashioned reproduction, adoption, surrogacy, and a huge range of infertility treatments.

If one accepts that the applications of reproductive science are driven by a utilitarian outlook, where there are tangible benefits to society and proportionately low risks, there is no future to reproductive cloning. However, I can envisage one scenario, outlined in the next chapter, where I think a baby born as a result of nuclear transfer could be justified. In that case, however, the baby would not be a clone of a person.

AFTER LOUISE

MAVERICK SCIENTISTS often claim that the work on Dolly and other animals has laid the groundwork for human cloning. I have two types of concern. First, as already discussed and well described by Stephen Levick, there would be unacceptable social consequences for a cloned child who was a genetically identical twin of someone else. Second, cloning research has raised serious concerns about the safety of both the birth mothers and the clones themselves, given the low efficiency and high risks of nuclear transfer. Although nuclear transfer is becoming safer and more efficient, the experience of cloning animals so far has shown that the possibility for harm is much greater than was the case with IVF, so much so that the benefits would have to be much more obvious in order for these risks to be justified.

To make Dolly we started with 277 reprogrammed eggs. Only twenty-nine made it to the stage where they could be implanted into thirteen surrogate mothers. Of those, only one became pregnant, carrying Dolly. Some claim that this success rate of 1 in 29 is not bad. They argue that reproductive cloning could even look attractive when it is remembered that these embryos have been screened for chromosome abnormalities and that this problem affects around half of the eggs of women aged under forty and higher percentages in older women.

But no reasonable and rational person would really want to clone a child given the dismal track record of my team and others worldwide. To repeat the Dolly experience in humans would mean obtaining around 300 eggs, which are already in short supply, persuading twenty-nine women to agree to having an embryo implanted. Of those, twenty-eight would risk the hurt and emotional turmoil of failed pregnancies, miscarriages, and deformed fetuses so that one embryo will "take" to produce a child. To have only one woman act as the clone's surrogate makes the whole enterprise even more horrendous. To me, this exercise in human misery and suffering seems inconceivable in a free society. Given the differences between man and sheep, the suffering required to create one human clone could well be much greater than this. Even a sinister billionaire in an unethical society would find it tough to indulge a narcissistic cloning fantasy. To clone humans is criminally irresponsible.

WILL HUMAN CLONES
HAVE BIG
BELLY BUTTONS?

SINCE THE BIRTH of Dolly there has been enough experience of animal cloning to build up a disturbing and detailed picture of the dangers posed by current techniques, above and beyond the psychological dangers. Independent groups have now published comprehensive reports on cloning cows that show safe reproductive cloning remains a distant prospect. One key research program in cloning has been led by Jean-Paul Renard at the Institut National de la Recherche Agronomique (INRA), at Jouy-en-Josas, near Paris. Renard was the first to clone a rat (notably Ralph the rat pup) and to clone a rabbit. His team also provided independent confirmation that cloning mice is possible and that cloning can be done by means of differentiated cells—the latter revealed when Marguerite the calf was unveiled at the Paris Agricultural Show (Vive le clone!, as *New Scientist* put it in 1998).

Renard's team has worked diligently to compare the pregnancies that occurred in around three hundred cows after the transfer of IVF embryos with those produced after transfer of cloned embryos derived from three sources of donor cells: adult cells, fetal cells, and cells from embryos at the thirty-two-cell (morula) stage. As one would expect from the checkered history of cloning, the more differentiated the donated cell, the less efficient nuclear transfer becomes. So when it comes to cloning an adult animal only 9 calves developed from 133 blastocysts, an efficiency of just under 7 percent. Many more embryos would have been produced, but failed to develop to the blastocyst stage.

In similar work David Wells and Björn Oback from AgResearch in New Zealand have found that from 988 somatic cell–cloned embryos transferred into recipient cows between 1997 and 2003, 133 calves were delivered at term, representing a 13 percent survival rate. Only 89 of those calves survived until weaning at three months of age, reducing the overall survival rate to 9 percent. This means that clones still have a success rate at least threefold lower than bovine IVF embryos, where about 30 percent of transferred embryos develop into healthy calves at weaning.

There is an unpleasant detail of this low efficiency that any human cloner should bear in mind. In most cloning experiments fewer than half of the embryos survive for even one week, by which time they should have reached the blastocyst stage. By contrast, almost all eggs fertilized after natural mating complete development to this stage. The deaths continue throughout the pregnancy of clones, unlike IVF, where all embryo losses take place by day 35, which is early in the 280-day pregnancy of a cow.

Because the loss of embryos produced by nuclear transfer continues throughout gestation, it would not be uncommon for more than 10 percent of cloned embryos that have developed successfully for the first third of pregnancy to die before the end of pregnancy (and double this percentage in the case of cloning using adult cells). This is true in all species that have been studied, regardless of laboratory, choice of donor cell, method of nuclear transfer, or anything else you can think of. The toll of miscarriage is high. Overall, the proportion of embryos that were born as calves in the two studies in France and New Zealand was fairly similar, at 7 percent and 13 percent, respectively.

The abnormalities that commonly occur during a clone pregnancy include overweight fetus or placenta, fluid accumulation within the placenta, particularly the allantois (the saclike membrane that develops into the umbilical cord and part of the placenta), inadequate attempts by the mother to give birth, and inability of the newborn animal to begin to breathe (it is unable to push blood through blood vessels that are many times bigger than normal). Less common abnormalities include failure of the immune system to develop and abnormal development of organs such as liver, kidney, and brain. I am confident that there are more horrors because, unsurprisingly, laboratories have been reluctant to describe all of the anomalies they have found.

The problems of cloning continue after birth: the viability of cloned animals is significantly lower than that of animals conceived naturally. In our first studies at the Roslin we found that four in every ten cloned lambs died within a few weeks of birth. This is a far higher proportion than would have been lost in commercial flocks in southern Scotland. In fact, the clones do even worse than these awful statistics suggest, since they had twenty-four-hour veterinary cover, unlike their peers in the average Scottish farm. Although we have learned how to take particular care of these unusual animals, the numbers that die just after birth remain significantly greater than normal. In the study by Oback and colleagues, referred to already, only 89 of the 133 calves that had been studied at the time of publication survived to weaning at about three months of age. This proportion, 67 percent, is far lower than would be expected for animals being monitored as intensively as the clones. Alarmingly, the groups in New Zealand found that the annual mortality rate in cloned cattle up to four years of age is at least 8 percent. This is in stark contrast to the long-term survival of the offspring of clones where mortality beyond weaning has thus far been nil.

Because of the technical shortcomings of many cloning experiments, it has been hard to quantify the effects, which vary among species. Often it has been impracticable to have an ideal group of animals with which to compare the clones, undermining attempts to evaluate the effects of nuclear transfer in a systematic way. Another problem is that, to boost the chance of success, surrogate mothers are loaded with more cloned fetuses than would result from natural mating, and sometimes this may cut their odds of survival if several begin to develop. (Renard's team in INRA reports that the survival of a single cloned embryo is about 7 percent, compared with about 10 percent when several embryos are transferred; this reveals that implanting more embryos does not produce the equivalent increase in the number of clones, because the death of one clone in utero can affect the development—and survival—of its brothers and sisters.)

Scientists also often prefer to move on and try something different than to dwell on a failure and why it happened. While understandable, this undermines the effort to figure out what is going wrong and why. Finally, there is the possibility that cloning produces abnormalities not seen before, and these could thus be overlooked by a pathologist who is looking only for what he expects to see.

There appear to be real differences in the precise pattern of abnormality between species and perhaps between different procedures and donor cell types. But the bottom line—cloning does not work very well—is the same regardless of laboratory, cloning technique, species, or choice of donor cell. In fact, the overall pattern of problems is so vivid that many of us have wondered whether there are any normal clones at all (in mice there do seem to be). Even in clones that are apparently healthy, the function of some genes lies outside the range that would be expected if the animals had been produced by normal reproduction. When the epigenetic mechanisms that regulate gene function go awry, the proteins that they describe are overproduced or underproduced. Perhaps a wider range of patterns of gene function is compatible with normal life than we had previously imagined. However, there may be long-term consequences that have yet to be understood: and they may emerge only when they affect something as complex as the human brain.

In the routine meetings between scientists at international conferences, other potential problems of cloning have emerged thanks to detailed studies. We owe the first description of a failure of the immune system in cloned calves to the efforts of Renard in INRA, who reported the death of a calf cloned from a cell taken from an ear of a calf, itself a clone of an embryo. The calf developed normally for the first six weeks but then suffered severe anemia: after birth, the number of white blood cells in the animal declined dramatically when it should have increased. They found atrophy of the organs in the lymphoid system, the thymus, spleen, and lymph nodes. The find was possible only because Renard and his colleagues had collected samples from all of their calves and, once the animal became sick, were able to go back and make a detailed analysis—diligence not often shown by other groups.

Clones can be left gasping for breath after birth, as happened with one cloned GM lamb born at the Roslin Institute. The lamb was the first to have a precise fragment of the gene that is involved in scrapie in sheep or BSE in cattle, known as PrP, removed, but there is no reason to think that this genetic manipulation contributed to its fate (BSE-resistant cows have now been cloned). At birth it was a fine, strong animal with a healthy appetite and bags of energy. It was separated from its mother to prevent any possibility of the spread of infection and was being hand-fed by a member of the staff of the Large Animal Unit, to whom it rapidly became attached. However, it panted all of the time, even when it did not exert itself.

None of the treatments that were recommended either by vets or by pediatricians were able to end its hyperventilation. We decided that it was kinder to end the lamb's life rather than to allow it to suffer further. Detailed studies by Dolly's pathologist, Susan Rhind, at the University of Edinburgh Veterinary School, revealed an abnormality in the blood supply to the lungs of the animal: there was a gross thickening of the muscle that surrounded small blood vessels close to the small air passages, bronchioles, in the lungs. Restriction of blood flow at this point would inevitably have made the lamb pant in order to take up oxygen and discharge carbon dioxide.

Another common problem with nuclear transfer is unusually big and heavy offspring—"large offspring syndrome." One can see it coming during pregnancy: gestation takes longer, and there is a sluggish onset of labor and a difficult birth. The larger offspring often fail to breathe after birth; less frequently, there are abnormalities of some organs such as the kidney and the brain. This syndrome had been seen before when an embryo was grown in less than ideal conditions, and it is now also linked, to a lesser extent, to any method used to handle and transfer embryos to a surrogate mother. It seems that the greater the number of procedures applied to an embryo, the greater the damage that is caused: even the handling of embryos causes abnormalities. Growing the embryo in a serum-enriched medium increases them markedly. (Serum was included in early embryo recipes because it was expected to be beneficial. We now know that while serum has some good effects, it actually does more harm than good.)

Even though the embryos of each species are affected to different extents, we already know of a human syndrome—Beckwith-Wiedemann—where size is increased. Because the process of IVF seems (the numbers are too small to be sure) to increase the incidence of this syndrome, it appears that reproductive technologies disturb the finely tuned mechanisms that regulate the way genetic instructions are carried out. However, the abnormalities occur most frequently and to the greatest extent in offspring produced by nuclear transfer: cloning wreaks the most havoc on these mechanisms of all. There have been claims that one day you will be able to spot clones on the beach by their large belly buttons, a relic of being born big with a big umbilical cord. That is, of course, if they survived that long.

SHORT-LIVED CLONES

MICE HAVE ALSO provided important clues to answer that all-important question of how long we should expect clones to live. They offer powerful insights because it is possible to clone large numbers, to compare and contrast different strains, and draw on a relatively huge number of related studies, for instance, on mouse embryonic stem cells. Experiments in the mouse are uniquely uniform, because in most experiments the mice and cloned mouse embryos are genetically consistent in a way that is impossible in any other species. Furthermore, the housing, environment, and diet are very closely regulated. Yet the results in mice are as variable and inconsistent as in any other species.

Only in mice has it been possible to describe in detail the effect that being a clone has upon life span. Because cloning methods are so recent, not enough clones of other species have had the opportunity to live out their natural life span. As one would expect, cloned mice do not tend to live as long as their naturally conceived peers, though there are now some instances of normal physiology, fertility, and life span.

To explain this result, some observers have focused on the effects of cloning on a cellular "timer," called a telomere, which consists of a strand of the genetic material DNA on the end of all chromosomes, rather like the plastic on the end of a shoelace. Every time a cell divides, the telomeres become shorter. One day, they eventually run out, and the cells in our bodies stop dividing. This cellular aging phenomenon is known as the Hayflick limit, after its discoverer, Leonard Hayflick, at the University of California, San Francisco. We found in the case of Dolly that, although we could turn back time in one sense, converting an adult cell into an embryonic cell, we did not reset the cellular aging clock, so that her telomeres were 40 percent shorter (older) than those of a typical sheep of her age. But this is not always the case, even in other sheep, and we wondered whether, given that we made the measurements on Dolly's immune cells, she was suffering from an infection that truncated her telomeres.

Later experiments have shown that Dolly was an exception and that telomeres are restored in most cloned animals, so cloning can be "rejuvenating" in this limited respect. The U.S. company Advanced Cell Technology described the cloning of six calves generated from fibroblast cells at the end of their life span, after 1,900 attempts. There were the usual problems of cloning, with individuals that were unusually large at birth and that suffered breathing problems and high blood pressure. Others appeared normal at two months. The telomeres of these animals all looked like those of newborn calves, ACT reported in 2001 in the journal *Science*, even though they were cloned from senescent cells where the telomeres have been worn down by aging.

But it is probably too simplistic to focus on telomeres alone when it comes to aging. An American team, led by Teruhiko Wakayama, cloned six successive generations of mice and found that telomere length increased slightly each time. But they lived only as long as other clones, underlining that the link between telomeres and aging is not clear-cut. Most likely, premature death was linked with a number of different health problems, such as severe pneumonia, necrosis of the liver, and cancer.

Cloned mice became obese in adulthood in one study. If this result extrapolates to people, a cloned human baby will risk growing unusually obese in middle age. Importantly, the offspring of these obese mice were all normal. Thus the problem seems to lie not in their genes but in the way that these genes were being used in the body: unlike genetic effects that are inherited by offspring, through the passage of genes, differences in the epigenetic mechanisms that regulate the functioning of the genes are not usually passed on to the next generation.

EPIGENETICS: SILENCE
OF THE GENES

CLONING SEEMS TO ALTER how genes are used, which is the focus of epigenetics—the study of changes in the ways that genes are used in the body that occur without changes in the genes themselves. Think of it this way: the genes that you inherit are all the possible protein "notes" that can be played by the instruments (cells) of the body. But each cell plays a different repertoire, depending on whether it is an eye cell or a brain cell. Not all genes present are functioning (expressed, as we like to say) in any one cell at any one time.

In the past few decades research has shown that common, complex diseases like cancer, diabetes, and heart disease stem from collections of changes in genes. This additional layer of complexity is reflected in the wide variation in severity, age of onset, ease of treatment, rate of progression, and other factors that affect diseases' variability. Epigenetics plays a role in cancer, for example. Before Dolly the sheep, this mechanism was also thought to explain why the cloning of adult mammals had never been accomplished: the simple tune played by an adult cell could not be wound back to the symphony of developmental possibilities on offer from an embryonic cell. In the wake of work by Rudolf Jaenisch at the Whitehead Institute for Biomedical Research and others, disruptions in the way genes are used in the body are now blamed for the deaths and deformities caused by nuclear transfer and other cloning techniques.

There are two ways in which epigenetic mechanisms act: one—called imprinting—is to make sure that, of the two copies of each gene in a developing embryo, only the mother's or the father's gene is used. The patterns of imprints are set during "gametogenesis," the development of the sperm or the eggs, which occurs a long time before fertilization. Imprinting is thought to govern the use of at least 75 genes, probably 100, and perhaps as many as 1,000. A clone will be healthy only if this fundamental imprint holds true.

Many of these imprinted genes are involved with regulating the size of the embryo—as discussed earlier, in the strange case of Kaguya, the mouse with two mothers. It was thought that nuclear transfer left imprints alone, given that they seem to be set in the ovaries and testes. But we at the Roslin and many other groups guessed, from the existence of large offspring syndrome, that cloning must disturb imprinting and thus the process that distinguishes maternal and paternal genes must be vulnerable to disruption in some way after fertilization. Later studies of cloned sheep by Lorraine Young (now at Nottingham University) and her collaborators demonstrated that one gene that is affected by nuclear transfer comes from the mother—a growth inhibitor called IGF2R, referring to how the gene makes a protein receptor that clears a paternal growth signal, called IGF2, from the circulation. As a result, the less IGF2R, the more IGF2 and the larger the offspring.

Recently, scientists from the United States, Sweden, and Japan implicated the same mechanism in tumor growth, where an "epigenetic" problem improperly turns on the copy of the IGF2 gene that should remain off: having a double dose of the IGF2 protein, which is the case in one in ten people, is sufficient to change the normal balance of cells within the lining of the colon, thereby doubling the risk that a cancer-causing genetic mutation will trigger a tumor there.

A second epigenetic mechanism gives each cell in the body its identity: using the same molecular mechanisms as imprinting, this ensures that, although every cell in the body has an entire genetic repertoire, only relevant genetic tunes are played: those required to make a skin cell, brain cell, or whatever. This "molecular memory," which enables a cell to remember what part of the genetic code to use to maintain its identity, is common to all creatures. That is why a brain cell stays a brain cell and does not turn into a liver cell, even though their complement of genes is the same. Because the mammary cell from Dolly's "mother" used only the relevant parts of the genetic code to make a mammary cell, it had been thought this would block the development of all the cell types required for a clone. Obviously, this memory was erased by the cloning technique employed to breed Dolly. But, of course, cloning is risky and inefficient. The reason is that the pattern of gene activity controlled by epigenetic mechanisms is not reproducibly wiped clean by cloning: thus you may not erase the entire "molecular memory" of its being a skin cell to obtain an embryonic cell. If this is faulty, some genes will be used inappropriately and others not at all.

A chemical mechanism called methylation is one kind of gene switch used in epigenetics: decorate a gene with methyl groups, and you turn it off. The cell no longer plays that "note." Normal development of an embryo is therefore dependent on the methylation state of the DNA contributed by the sperm and egg, the parental imprints, and on the changes in methylation of DNA that help direct an embryonic stem cell to turn into a skin cell or whatever. The workings of cells are, to be sure, vastly more complicated than this.

Some scientists even suggest that the genetic code of DNA is joined by a second code, called the histone code, named after proteins called histones, which govern the way that the body interprets genes. It could well be that understanding the nuances of this mysterious code, in addition to imprinting, can shed important light on what goes wrong in cloning.

The histone code relies on how the 2.3 meters or so of DNA in each cell is bundled up within chromosomes and, more important, how it unpacks. Here, the popular picture of chromosomes is muddled and misleading. The X shape used to depict chromosomes in many accounts actually shows the chromosomes during the process of cell division, when they are copying themselves. For the rest of the time, the chromosome would look more like a sausage with a pinch in the middle. But perhaps it is better to forget chromosomes altogether, given that we are most interested in what happens when the cell is reading the genetic notes on the coils of a single DNA molecule bundled into a chromosome: if you can see the chromosomes, then the cell is going through cell division and the DNA wrapped up inside them is not doing much more than contributing to this process. Only when a chromosome unravels to leave a spaghetti-like mess of DNA in the nucleus of a cell does the serious business of playing a genetic tune begin.

The raveling and unraveling of DNA is controlled by proteins. In fact, chromosomes are made of a blend of the double helix of DNA and protein that together are called chromatin. In one form, dubbed heterochromatin, the genes are all packed away and put out of use. This is the stuff that the visible regions of the chromosomes are made of. In the second form, called euchromatin, the DNA spills out into the interior of the nucleus so that the genes can be put to use—played, to continue the musical metaphor—to make proteins that can build and operate the cell.

A conductor's-eye view of a cell can reveal that, in the apparent chaos of this DNA mess, there is order. One would expect that when the chromosomes unpack so that their genes can be used, the nucleus of a cell is filled with a tangled mass of DNA from all the chromosomes (forty-six in the case of a human). Instead, you discover that the chromosomes are always found in the same territories in the nucleus, even after the process of spaghettification. Jane Skok of University College London can study the position of the genes encoding antibodies in certain white blood cells (called B cells) by means of DNA probes that are able to bind to them. By analyzing the position of these genes in individual cells, she is able to see that in a cell where a gene is being used, the DNA that describes it is somewhere in the middle of the nucleus. When the gene is out of action, it is tucked away on the periphery of the nucleus.

Of the various proteins that control the way the message of DNA is read, perhaps the most important are the histones. Certainly they are the best understood. Think of a cheap string of pearls, where each pearl is a spool around which DNA is wound and the elastic is the DNA. Around a ball of eight histones the DNA is wrapped two and a half times to form a "pearl," called the nucleosome. The cell "reads" the DNA's message rather in the manner of groping for the pieces of elastic between the histone pearls. In a chromosome the nucleosomes are packed tightly together, like a necklace that has been wound up so that it can be easily put into a purse.

To control the packing and unpacking, and thus the way that genes are used, the histones in each nucleosome have a "tail," dangling into the nuclear soup, like a drawstring for the necklace. Chemical modification of this tail controls the way genes are used: changes can separate the pearls to expose DNA strands between them, move them around the DNA, or pack them together and so lock genes away. If a process called histone acetylation takes place, genes tend to go into action: the nucleosomes move apart to allow the DNA to unravel so that certain proteins in the soup of molecules swimming around in the nucleus can go to work on the genes that link the loosened nucleosomes and carry out their instructions. Histone methylation (not to be confused with DNA methylation) acts in two ways; depending on where the methylation mark is positioned, it can turn genes either off or on.

The histone code is crucial for development. Compare a body cell (somatic cell) with a germ cell, such as a sperm, and the chromatin structure is quite different. In the few hours after a reprogrammed embryo is activated, important changes to chromatin take place. At least one histone has a form that appears only in the embryo. All this suggests, as one would expect, that in the turning of a spherical fertilized egg into a fetus, there are lots of changes to gene regulation. Meanwhile, studies of cloned mice show that changes to chromatin as well as changes to DNA methylation can be messed up by nuclear transfer. And, as a result of this "nuclear remodeling," some genes are switched on when they should be switch off, and vice versa. The embryo dies. The fetus fails to develop normally. An offspring is born with serious abnormalities.

To prevent these problems we need to know more about the way genetic material is used in the first, critical twenty-four hours after an egg is fertilized. This is now much more the focus of my interest than nuclear transfer itself. It should already be apparent that the regulation of genes is highly dynamic and complex: it is going to take a long time to understand these epigenetic mechanisms, to know what normally happens, and then intervene during cloning to fix what nuclear transfer fails to achieve. At present we expect too much of eggs in reprogramming adult DNA and have to give nature a helping hand. We need to improve nuclear transfer.

BEYOND DOLLY

Efforts are under way to hone nuclear transfer in various ways. The Dolly method involves highly precise methods to remove and inject DNA under a microscope, using manipulation apparatus worth tens of thousands of pounds. Of the many variations on this theme that have since emerged, I have come to prefer two quicker and easier alternatives, called zona-free methods, which are less reliant on expensive equipment. One was developed by Teija Peura and colleagues from the South Australian Research and Development Institute in Melbourne, and refined by Gabor Vajta and his colleagues in Denmark. The second was developed by Björn Oback and colleagues at New Zealand–based AgResearch, and the University of Waikato. In the first zona-free method, Peura uses the following shortcut in enucleating the egg. Rather than hunt the chromosomes and remove them specifically, the egg is simply cut in half with a handheld blade. Cutting in half was the approach to enucleation used by Steen Willadsen, as was discussed in chapter 2. Ultraviolet light is then used to identify and discard the half that contains the chromosomes. Two halves of eggs emptied this way are put back together again, like Humpty Dumpty. Nuclear transfer is achieved by fusion of the donor cell with this reconstructed egg. Although it takes two eggs to make each clone, it is likely that the reconstructed embryo will be slightly larger in volume than those that resulted from the Dolly procedure and thus have more cellular machinery in the cytoplasm to draw on and therefore a better chance of developing well. One potential problem with this method is that the resulting embryo contains mitochondrial DNA from two different eggs.

The second zona-free method developed by Oback uses micromanipulators in a way that is much quicker to learn and to carry out. However, it does not offer any increase in efficiency. Enzymes are used to strip the egg of its zona pellucida. The naked egg is then held against a pipette, rather than being sucked against it, as in our procedure. Ultraviolet light is used to guide enucleation with another pipette, a process so quick that there is little damage. Then a donor cell is positioned next to the naked egg in the presence of lectins, sticky compounds that glue the two together, and the two fused with an electric pulse. In the Dolly method the egg and tiny donor cell had to be aligned exactly by hand between the electrodes—so that the membranes at the point where they touch are perpendicular to the field—for fusion to occur. This alignment is essential if the two cells are to fuse efficiently. Strikingly, in this zona-free method the electric field does all the hard work and lines the two up properly—the egg and donor cell roll around to align so long as the zona is not there. This technique takes an inexperienced person a few weeks to learn, compared with six months for the original Dolly method. Even though this method does not increase cloning efficiency, it effectively doubles the throughput in cloned embryo and cloned offspring production. Because of such advances, we can see the day when robots and computers will carry out the delicate manipulations required for cloning.

FLASH OF INSPIRATION

ANOTHER FACTOR that could make cloning safer is greater ease in achieving activation, the moment that a flash of calcium triggers development. When this happens naturally, one can see pulses of high calcium concentrations every few minutes that continue for several hours. There are various ways to trigger a surge of calcium in cloning, for instance, with an electrical current or even giving the reconstructed eggs a calcium bath. But most generate one very large wave of high concentration calcium for hours. The development of successful methods for nuclear transfer in a different species has involved the adaptation of the method of activation for that species. It is crucial to get the details right to be successful.

To improve cloning in general it would be good to try to mimic the natural waxing and waning of calcium after fertilization. A group in France led by Jean-Pierre Ozil has spent years working on a box of electronics to create precisely the same sequences of calcium pulses that occur during fertilization, giving eggs bursts of calcium in a tiny bath. This has shown that it is possible to increase the success rate of producing parthenotes, but it has not yet been incorporated into routine cloning procedures, because it is still a lab prototype and not yet on the market.

An alternative strategy is to find out more about how activation occurs naturally and to identify and use the particular protein in the sperm that triggers development of a fertilized egg. The search for the sperm signal that awakens the egg began two centuries ago and is vital for understanding how a fertilized egg starts life anew. Recently, this trigger was revealed to be a protein—PLC-zeta—by Karl Swann of University College London and Tony Lai at the University of Wales College of Medicine. This protein switch could offer a sophisticated alternative to currently used activation methods.

CYTOPLASMIC ENGINEERING

AN ALTERNATIVE WAY to get development—from imprinting to the histone code—off to the right start in nuclear transfer is to pretreat the donor nuclear DNA in some way. One approach, developed by John Gurdon, is to have faith in how nature uses very similar machinery to do important things in a cell, whatever the species. At a particular stage in frog egg development (frog eggs offer plentiful cytoplasm compared with tiny mammalian eggs), he injected in a mammalian cell that had previously had holes in it, so that it could set up a "chemical dialogue" with the frog cytoplasm. Through this experiment, done in the 1970s, he could tell that some of the mammalian genes were turned off and some egg genes were used. More recently, he has shown that one (Oct4), known to signify embryonic stem cells, goes into action. In collaboration with colleagues in the institute and Julian Blow of the University of Dundee, we are investigating how a similar approach could help reprogram the DNA used in nuclear transfer.

Complementing this is the use of technological advances in genomics to screen the expression patterns of tens of thousands of genes to identify developmental differences between "reconstructed embryos" and those produced in vivo or by IVF. This study will be crucial for the efforts to see how assisted reproduction techniques can make genetic melodies played by each cell drift off key or off tempo to disturb development.

There is another interesting feature of the first twenty-four hours of life. As was pointed out before, the parental DNA has not even been mixed to form an individual at this stage, and the DNA is not even being used (the details of how RNA and DNA are used in the zygote vary from species to species), as it is in a normal cell. We need to know how parental DNA is organized at this critical time. And we have to know about the parental genetic messages scurrying hither and thither to control and influence the early stages of development. These messages are written in the code RNA and are bequeathed with proteins to the zygote (early embryo/placenta) mainly by the egg, but a small number are contributed by the sperm. One reason that MII eggs work is that they possess the right RNA instructions to help direct protein manufacture and development, perhaps also organize the chromatin. Although their exact function is unknown, it is the RNA that guides the development of the early embryo, before it starts to use the instructions in its own genetic code—just as you need an operating system to run a program on your computer. One can think of this as an RNA operating system that enables the development software to run in the cell.

One experiment that shed light on this hitherto unknown aspect of our inheritance was performed by Stephen Krawetz of Wayne State University School of Medicine, in Detroit, and colleagues in Britain, which revealed that men play a more important role in procreation than they may have thought. Despite cloning and Kaguya, the first fatherless mouse, Krawetz found that sperm do more than simply deliver a father's DNA to an egg. There were six messenger RNAs, each corresponding to a paternal gene, in human sperm and in fertilized eggs that were not in unfertilized eggs. This suggests that sperm deliver at least six RNAs—probably more—to the egg at fertilization. Krawetz has also found micro RNAs, a recently discovered large class of regulatory, noncoding genes that bind to sites in target messenger RNA to regulate how they are broken down and used in the zygote. In humans sperm also deliver a structure called a centriole, a key part of the machinery of cell division. These findings could have implications for the success of cloning, where only DNA is transferred. Perhaps, by adding a sprinkle of paternal RNA as well, the low success rate could be boosted for research to clone embryos as a source of stem cells for treatments.

Another focus of efforts to improve on the Roslin technique rests on a feature of Dolly that is not widely appreciated: although the process of cloning exchanges the nuclear DNA of an egg for that from a donor, it leaves mitochondrial DNA from the egg. Thus Dolly had the nuclear DNA from a mammary cell from a Finn-Dorset sheep and the mitochondria of the egg, which came from a Scottish Blackface. Strictly speaking, Dolly was not a clone because of this difference in mitochondrial DNA, but most of us will continue to use the familiar term when referring to her.

Dolly therefore relied on genetic material from the Blackface—around twenty genes in the donor egg's mitochondria. While the amount of information in the mitochondria is small in comparison with that in the chromosomes, it is essential for normal development. Abnormalities in mitochondria, which can accumulate with aging, can lead to disease. That goes for human clones, too, where in most cases the egg that was used during nuclear transfer will have been obtained from a different woman from the one who donated the DNA, if the latter is a woman at all. The relatively low expectations we have of a fertilized egg—whether in the laboratory or in the womb—reflects the important differences between eggs in their ability to support normal development of an embryo. These differences in the success of development can be due to differences in the mitochondria and may mean the difference between the life and the death of an embryo. We need to understand more about what causes them and how to tinker with these factors if we are to improve success rates.

THE FOLLY OF
REPRODUCTIVE CLONING

THERE ARE SOME PEOPLE, of course, who want us to go further and take cloning to term. They have an easy answer for the difficult problems I outlined above: simply select the good embryos where there is no evidence that gene regulation has gone awry. This suggestion rests in a misleading way upon the fact that it is already possible to identify those embryos that have inherited known genetic abnormalities from a parent, thanks to a technique called preimplantation genetic diagnosis. In the case of cloning, however, we are looking not at faulty genes but at faults in the way genes are used, whether due to errors in the histone code, methylation, or whatever. And to make matters worse, we don't understand the details. In fact, our understanding of the subtleties of gene regulation is pitiful. One cannot devise a test for a poorly understood problem.

To try to clone a baby using current methods would be a terrible gamble. Imagine tossing five coins and getting five heads or five tails. That is an unlikely outcome, and yet for a clone to survive and be normal would be about as likely as seeing five heads. It seems self-evident that it would be grossly irresponsible even to consider using the present procedures to produce a cloned child. The likely outcomes of any such attempt would include late abortions, the birth of dead children, and, perhaps worst of all, the birth of some children that survive, but who suffer a lifetime of pain and disability. Remember the cloned lamb that panted all of the time because of an abnormality in the lungs. It is distressing enough when an animal suffers this way, but it would be awful to have inflicted this fate upon a child.

One is tempted to conclude from this that reproductive cloning should never be tried until proved totally safe. But, as we all know, absolute safety is impossible, and if we used this simplistic rule we would still be living in the dark ages. Yet it is true to say that, for now at least, reproductive cloning must be banned because it is too risky. The problems look daunting. But I would never go so far as to say they will never be overcome. Everything rests on details of the technique, such as the use of fresh eggs. I have no doubt that revealing such details with new studies, and with techniques that will attempt to show whether cloned embryonic stem cells are normal, will help show how to deal with many of the problems facing cloning to produce children. Even when safe reproductive cloning is possible, I believe that there will not be any obvious application, because there will be alternative ways to meet apparent needs. Specifically, I believe that it will be possible to produce gametes for infertile individuals. Several groups have now shown how embryonic stem cells can develop in the laboratory into the early forms of cells that eventually become eggs or sperm. If the stem cells were derived from a cloned embryo, then the gametes could be used by the person from whom the nuclear donor cell was taken.

In early work with mouse embryonic stem cells, Hans Scholer's team at the University of Pennsylvania used embryonic stem cells to make cells with some of the characteristics of eggs. Already, Toshiaki Noce, of the Mitsubishi Kagaku Institute of Life Sciences, in Tokyo, has pushed mouse embryonic stem cells toward sperm production. George Daly and colleagues at the Children's Hospital Boston/Harvard Medical School and the Whitehead Institute for Biomedical Research have injected immature mouse sperm created this way into eggs to make embryos.

Their work also opens up the possibility, one day, that eggs and sperm could be grown from stem cells and used to study the effects of hormone-disrupting chemicals, for assisted reproduction, therapeutic cloning, and the creation of more stem cells for further research and for the improved treatments for patients suffering from a range of diseases.

Though it remains to be seen which of these more speculative strategies will eventually pay off, these techniques offer many new opportunities for infertile couples to reconstruct their eggs and sperm to have children that are genetically their own, that will contain a blend of their genes, as would normally result from making a baby. Cloning, meanwhile, offers only the unusual option of having an identical, and younger, twin of one parent. In short, I feel confident that there will be many alternatives to the use of reproductive cloning. It will not be necessary. Nor do I feel that society will ever be comfortable with reproductive cloning. Attempts to clone offspring will be condemned as unethical.

BEYOND HUMAN CLONING

LIKE MANY of my colleagues, I believe that the prospect of routinely growing cloned embryos so that cells and tissues can be grown for patients is unlikely on a large scale or in the long term, though crucial in the short term. Crucial because we know that this class of cells, above all others (notably adult stem cells), has the wherewithal to develop into any type. If we can perfect this process and make it reliable, then we will not only obtain cells for repair and for testing drugs but also obtain profound insights into the details of development and disease. And on the basis of information from this research on therapeutic cloning, and insights from the study of more differentiated "adult" stem cells, we can devise embryo-free methods that can wind back the developmental clock of a patient's cell so that it can recover its full genetic potential. If we can understand what happens in the wake of nuclear transfer, we will one day be able to mimic those events and thereby actually change cells from one type to another—so that after a heart attack, say, skin cells from the patient might be turned by a cocktail of factors into heart muscle cells and used to repair the damage to the heart. Thus, immunologically matched cells would be derived without the expense and difficulty of cloning. We would be able to grow cells of any type to repair a patient without the need to create human embryos—and, of course, without inviting the disgust of those who still believe that life begins at conception.

There are already hints that this feat of cellular alchemy is possible. In 2002 Philippe Collas of the Institute of Medical Biochemistry, University of Oslo, reported how he gave one type of body cell—fibroblasts—some of the characteristics of another type, white blood cells called T cells. His team made holes in the cell membranes of the fibroblasts and dunked them into an extract from the T cells. This "reprogrammed" the fibroblasts to lose some fibroblast functions, but to gain some from the T cells. Since then he has converted fibroblasts into insulin-producing cells, albeit transiently, and fat cells into the cardiomyocytes of heart muscle.

Another approach has been taken by Peter Schultz and Sheng Ding at the Scripps Research Institute in La Jolla, California. They have been screening molecules in a hunt for those that can turn the clock forwards or backwards in a cell's development. The chemists are on the trail of a small molecule they call "reversine," which can cause a muscle cell to turn into a progenitor cell that has more developmental potential and could be turned into a wider repertoire of tissue types.

Ultimately, scientists want to use these methods to persuade body cells to "dedifferentiate"—make the journey back to earlier, more plastic stages. Thus cells of any type could be regenerated from a speck of a patient's tissue for a vast range of treatments and repair. This approach would satisfy the main promise of therapeutic cloning: create cell lines tailored to a patient that avoid the problems of immune incompatibility. It would also eliminate the need to obtain donated human eggs and, most important, avoid the moral outrage of those who oppose the use of human embryos.

However, one other application of nuclear transfer still involves embryos. I think that it falls between the extremes of reproductive cloning and therapeutic cloning: this clone is not of a person but of a blastocyst. And, unlike reproductive cloning, this combination of genetic modification and cloning seems to me to offer an important way to reduce human misery and therefore deserves consideration.

EIGHT

DESIGNER
BABIES

POLLY THE SHEEP has something important to say about the future of humanity. The combination of nuclear transfer and genetic modification used to create her have much broader applications in human medicine that could prevent a great deal of suffering. They may help to prevent the birth of children with devastating genetic diseases. Because public attitudes have shifted from alarm to acceptance in the case of IVF, I can even anticipate the day when the genetic modification of human embryos is accepted as routine, even if it is not used on a large scale. When it is possible to produce genetically modified people safely and predictably, which is still a distant prospect, mankind will take on more responsibility for his own evolution.

But I can see profound ethical issues arising from what some condemn as "playing God." Patients, parents, and regulators will bear a heavy burden of responsibility when it comes to deciding whether this technology should be used and, if so, in what circumstances. Some believe that the desire to prevent the birth of disabled children is often driven by misinformation and prejudice. Others see this as another unwarranted intrusion into natural procreation. Many fear the subtle long-term effects on society: Would a child's sense of independence or self-worth change on learning that his parents had intervened in his birth to change him in some way? These technologies could also be used to make someone not just well but better than well, to expand muscles, enhance memory, and delay aging.

I am not working on this technique—at least not at the moment—but consider it important that the opportunity is discussed because of its potential impact, from benefits to pitfalls, and the difficult choices that lie ahead for individuals as well as society. However, there is much confusion about exactly what scientists propose to do, what they are actually capable of doing, and whether they should be allowed to do it at all. Several techniques are already on offer to deal with hereditary disease, such as gene transplants, better known as gene therapy.

GENE THERAPY

THE MOST advanced technology that we possess to tackle the root cause of genetic disease is gene therapy—gene transplants, if you like. This therapy can, for example, be used to correct the damaged genes responsible for diseases such as sickle-cell anemia. Sickle-cell disorders are most common in individuals of African, Mediterranean, Indian, and Middle Eastern descent, and one in every thirteen African Americans carries the sickle-cell trait. The disease is caused by a single genetic "spelling mistake"—a single nucleotide mutation—in the human beta globin gene that leads to an abnormal form of hemoglobin, a protein that can capture and release oxygen.

One copy of the sickle-cell gene is beneficial, providing an increasing amount of protection against malaria during the first decade of a child's life, according to a study by Tom Williams of the Kenya Medical Research Institute/Wellcome Trust Research Program in Kilifi. However, children who inherit a double dose of the sickle gene suffer from sickle-cell disease, which in its worst form causes chronic ill health and can result in early death.

Red blood cells supply oxygen to the body by means of their cargo of hemoglobin that normally floats freely within them. But sickle-cell patients have a mutated, or "sticky," form of hemoglobin that tends to clump together into long fibers. The fibers form a scaffolding that distorts the cells into their namesake "sickle" shape, so they clog small blood vessels. These traffic jams deprive vital organs of oxygen, and patients end up with anemia, jaundice, major organ damage, and many other maladies. Theoretically, gene therapy could cure this disease.

To do this, scientists take a corrective gene—a normal form of the beta globin gene—and introduce it into the bone marrow cells of patients. The job of genetic surgery can be done by a modified virus, a complex chemical that has been honed by nature to insert its genetic code into cells and pirate their cellular machinery. Rather than convert a cell into a virus factory, the virus would be modified so that it inserted a therapeutic gene. Another approach is to package the gene in a fat particle, called a liposome, or even to inject naked DNA.

Gene therapy has been tested many times. The first approved human trial of gene therapy was carried out by Michael Blaese and colleagues in September 1990 at the National Institutes of Health in Bethesda, Maryland. At that time the media hailed this experiment as the dawn of a medical revolution that would be able to treat the four thousand or so hereditary diseases caused by a defect in a single gene, even cancer, heart disease, and other ailments where genes play an important role. But because the two young girls who received the gene therapy—sufferers of a rare immune deficiency—also continued to receive conventional treatment, the results of that pioneering trial were ambiguous.

Although animal studies continued to suggest that gene therapy had great promise, human trials during that decade were disappointing. A glimpse of the true potential emerged only in 2000 when gene therapy produced truly remarkable results. A paper in the journal *Science* described how the therapy had released three boys from imprisonment in a germ-free jail. Alain Fischer of the Hôpital Necker in Paris and colleagues treated the baby boys, aged between six and twenty-one months, affected by a genetic disorder that causes a severe immune deficiency that strips their bodies of protection. Called severe combined immunodeficiency (SCID) X1, the X referring to how it was inherited on the X chromosome, the disease forced the boys to live within sterile "bubbles" (children born with acute SCID are known as bubble babies). In this way they could avoid any threats, for instance, from a cold sore or chickenpox. They would die within a year without a bone marrow transplant to rebuild their nonexistent immune system, which works in only around 60 percent of the patients and necessitates additional treatments.

Instead, thanks to the gene therapy, the boys were given a normal copy of the defective gene responsible for the disease. First their bone marrow was removed. Then the stem cells in the marrow that are the "parents" of the cells that form the immune system were enriched, infected with a mouse virus containing the corrective gene, and finally returned to the body. The lives of these boys were changed out of all recognition by this genetically altered mouse virus.

Previously, the boys had been affected with severe diarrhea and unable to eat because of chronic gut infection, so they had to be fed intravenously. After gene therapy they had a normal gut and could eat a diet conventional for their age. No longer in protective isolation, they were able to live at home without any treatment, enjoying normal growth and development. This transformation was remarkable, and success was also reported in similar trials by Adrian Thrasher and Bobby Gaspar at Great Ormond Street Hospital, London.

But the brilliant record of the French team would become tarnished. Almost three years after the treatment, white blood cells (called T cells) of the two youngest patients began proliferating abnormally, as they would in the blood cancer leukemia. Subsequent investigation showed that the French vector (the virus) had turned on an oncogene—a cancer-causing gene—called LMO2, which is required for the production of blood cells. One child died, and, at the time of writing, the third had developed leukemia. Despite this serious setback, the British trial has not suffered these severe side effects. Gene therapy is still thought to offer a safer alternative to bone marrow transplants from a parent or mismatched unrelated donor. And the vectors are being made safer and more effective all the time.

Aside from the dangers, which can arise in any experimental technique, another consequence of this kind of gene therapy should be highlighted. The direct influence of the transplanted gene is confined to the boys alone. The genetic repair carried out by the French and British teams would not be inherited by any children that the boys might subsequently father, because the gene transplant affected only the bone marrow. This is so-called somatic gene therapy— where somatic cells are nonreproductive cells—and this limitation is crucial for ethical approval, since there was, and remains, unease about using "germ line" gene therapy, which would pass the new gene on to a recipient's descendants. The human gene pool would be permanently affected as a result. Some even talk of their creating "genetic pollution." Although these changes would presumably be for the better, the fear is that an error in technology or judgment could have far-reaching consequences.

THE STORY OF ANDI

GENE THERAPY can also be carried out at a much earlier stage of life to affect all cells in the body, including germ cells. This was attempted in the creation of the world's first genetically modified primate, a baby rhesus monkey called ANDi ("inserted DNA" spelled backwards). Gerald Schatten, then at the Oregon Regional Primate Research Center in Portland, led the group that added an extra gene, one that codes for jellyfish green fluorescent protein. This protein was of no medical interest but was very easy to detect—it makes the recipient glow green under ultraviolet light. In this way the success of ANDi would give the green light, as it were, for the creation of primate models of awful diseases, such as Alzheimer's.

Leaving aside the vexed issue of whether primates should be "humanized" for animal research, the risks taken in creating ANDi would be unacceptable for use on people. Schatten and Anthony Chan used a virus containing the gene in the form of RNA code to genetically modify 224 eggs and then fertilized them to produce forty embryos. After the embryos were transferred to twenty surrogates, only five pregnancies resulted. Three healthy male babies were born, two infants were stillborn, and another pregnancy did not develop. Of the healthy infants, only ANDi contained the new marker gene—revealed when the team amplified and analyzed genetic material extracted from the inside of his cheek, his hair, and cells in his urine, as well as studies of his placenta and birth cord. Even so, although ANDi took up the gene, the experiment was not quite a glowing success: the gene did not seem to be in use in ANDi's cells.

The experience of the French team with immunocompromised children and of the Americans with ANDi seems very familiar to someone like me who spent years altering the genetic makeup of mice, sheep, and cattle. We have all found that it is possible to make genetic changes, but there are serious limitations. First, it is only possible to add genes. Second, the efficiency is so low that only a small proportion of the offspring carry the additional gene. Finally, the genes are not always used in the body (expressed, as scientists say). And, as the French team showed, even if new genes are successfully taken up, they can cause problems. Quite clearly this is a technique that has to be greatly refined before it can be put to routine use in human medicine. However, an alternative way to deal with hereditary disease that does not require genetic tinkering is already in widespread use.

SELECTING EMBRYOS

IN THE OLD DAYS, couples who were aware that they had a history of a genetic disease in their families had no option but to take a chance and have children not knowing whether any or all of them would inherit the disease. For them life was a lottery with very high stakes. Take the example of hemophilia A, the most prevalent and serious type of hemophilia. Unlike Christmas disease, which we encountered in chapter 4, this more common form of the disorder is caused by a mutation in the gene responsible for factor VIII, a protein made mainly in the liver that plays a role in clotting. The mutation can arise spontaneously, as it did with Queen Victoria, who reigned over Great Britain between 1837 and 1901.

This tiny DNA alteration sent ripples through history. Her eighth child, Leopold, was affected and died aged thirty-one from a hemorrhage. Two of her daughters, Alice and Beatrice, were carriers and passed the faulty gene to several royal families in Europe, including in Spain and Russia. In the latter case, the gene passed on to her granddaughter, Princess Alexandria. The princess was married to Czar Nicholas of Russia, and their son, the czarevitch, was also affected with the disease. The princess and the monk Rasputin, who treated the czarevitch, had considerable political influence, and this faulty gene may have been responsible in part for precipitating the Russian revolution.

One step towards ending this kind of deadly lottery came with the development of various blood tests that could help check the health of the unborn child. For example, there is a blood test that can be used by doctors to estimate the risk of the fetus's having certain defects, including spina bifida, Down syndrome, and other chromosomal abnormalities or structural defects. In the case of Down syndrome, caused when the fetus has an extra copy of chromosome 21 and thus an overdose of the 225 genes that it carries, levels of two particular proteins in the blood of pregnant women have been found to be an indicator of risk. The unborn baby is considered at high risk of the syndrome if free beta human chorionic gonadotropin, a hormone, is present in high concentrations, and pregnancy-associated plasma protein A in low concentrations. Fetuses with Down syndrome also tend to have an excess of skin at the back of the neck (nuchal translucency), which can now be seen with an ultrasound scan.

To make certain of their diagnosis, doctors use amniocentesis and chorionic villus sampling (CVS), both of which rely on obtaining fetal cells so that their chromosomes can be analyzed for defects. Amniocentesis has traditionally been performed at fifteen to eighteen weeks gestation, when a needle is used to collect fluid from the amniotic sac. CVS can be done as early as ten to twelve weeks gestation, when a sample is taken from the mat of chorionic villi, or fingerlike projections on the fetal side of the chorion frondosum, the early placenta. Both carry a risk of miscarriage, so they are usually used only when blood tests suggest there may be a problem.

Alternative tests, which exploit the discovery that genetic material from the baby, such as messenger RNA, circulates in the bloodstream of pregnant women, offer the prospect of a safer and earlier way to screen the fetus for genetic disorders. Whatever the test, however, if it reveals that damaged genes are present, then prospective parents are faced with the choice of either continuing with the pregnancy to produce a child known to have inherited the genetic disease or terminating the pregnancy. A dreadful dilemma.

New hope for families with a history of genetic disease came in 1990, when Robert Winston, Alan Handyside, and colleagues from Hammersmith Hospital in west London announced that they had developed a way to test IVF embryos so that only those that were free of a serious defect were implanted in the mother. They carried out a genetic test on DNA amplified from a single cell taken by a needle from the early embryo, when it consists of a ball of six or eight cells and is only a tenth of a millimeter across. Fears that this biopsy would be likely to increase the risk that this embryo will not go to term do not seem to have been founded: pregnancy rates after this method of preimplantation genetic diagnosis, or PGD, are the same as for normal IVF.

The team initially used this method to screen the sex of embryos from parents who carry disorders that would affect only boys, such as Duchenne muscular dystrophy, a fatal cause of muscle wasting, and childhood adrenoleukodystrophy, a devastating progressive disease that is best known because of Lorenzo Odone, whose story is told in the 1993 film *Lorenzo's Oil*. That way they could ensure that only female—and thus healthy—test-tube babies were transferred to the mother. Winston said at the time, "For the first time, people who are carrying a severe genetic disease can start a pregnancy from the beginning, knowing it is normal, and do not have the specter of a termination of pregnancy."

Since then PGD has been extended to the analysis of embryo cells and to that of the polar bodies cast off either before or after fertilization and broadened to many other hereditary diseases, to screen out genes that raise cancer risk, to rhesus factor disease—the potentially fatal condition caused by incompatibility between a baby's blood and that of its mother. The technique will soon be able to screen as easily for any defect as conventional methods, such as amniocentesis, thanks to new methods to amplify all the DNA in the cells taken in the biopsy, rather than the DNA linked with one specific disease, a method called whole genome amplification. The method can also be used to check the number of chromosomes, since chromosomal abnormalities play a major role in miscarriage.

The latter method, "full chromosomal analysis," was developed by Dagan Wells and Joy Delhanty at University College London Medical School. The technique combines whole genome amplification and comparative genomic hybridization, which detects chromosomal imbalances, to assess the copy number of every single chromosome in the majority of cells of a preimplantation embryo. Strikingly, the technique revealed that most human embryos contain a proportion of abnormal cells, some of which arise in the egg or sperm and others from early development. The result is a mosaic of normal and abnormal cells, which is found in 60 percent of embryos by the age of three days. For those who hanker after a "normal" or "natural" birth, this should give some pause for thought. In terms of numbers a "normal" embryo has chromosomally abnormal cells.

As ever, PGD raises a number of issues. First, there are questions about the safety of the IVF and embryo manipulation required for PGD. Although hundreds of babies have been born after embryo biopsy with no obvious harmful effects, there needs to be a more systematic follow-up of their health over decades to be sure. There are also uncertainties about the accuracy of the genetic tests that are used. Some find PGD, or its uses, morally unacceptable because it involves the creation and selection of human embryos. There are issues about whether it should be used to select against susceptibility as well as disease, and whether it is right to choose a child's sex or other desirable characteristics (though attributes such as musical ability and intelligence are so genetically complex they would seem beyond this method). Other issues have to do with whether a couple who are blind or deaf, for example, can use PGD to select an embryo with a matching disability. Questions of equity arise: many cannot afford PGD, and it is not clear whether and to what extent health insurers and national health systems will cover the costs. There are also broad ethical questions about the impact of PGD on family relationships, people living with disabilities, and society as a whole. For what it is worth, I feel that, under the kinds of control we have seen in Britain, and with proper advice given to parents who want to undergo PGD, the technology is very unlikely to cause harm and it offers many benefits. It certainly provides a kinder alternative to getting pregnant, having amniocentesis, and, if positive for a serious disease, having an abortion.

Most controversially, the technique has been used to create what pro-life campaigners call a "designer baby" to signal their disdain. The pejorative term was presumably meant to give the misleading impression that parents can now choose embryos as casually as they select a piece of designer clothing. But to me embryo selection does not count as embryo design. Nor does it seem very different in principle from the selection of healthy-looking embryos for IVF. In one highly publicized case, campaigners challenged the authority of the HFEA to license the use of the method to create a "savior sibling," a healthy embryo with the correct genetic match to allow cells from its umbilical cord to be used to treat an older sibling. In this case the older child had the serious blood disorder thalassemia, and the aim of the exercise was to restore his supply of normal red blood cells. Approval was eventually given, after a court battle. However, even after the family had overcome these formidable legal obstacles, no one could guarantee success. In the first attempt fifteen embryos were produced. Only one proved to have an exact tissue match, but it carried the beta thalassemia disease. In the second, ten embryos were produced. Two of these proved disease free and to have a tissue match. One was implanted, but no pregnancy resulted.

Creating a "designer baby" is difficult. There is a limit on the number of eggs that can be produced at a time by a woman. Not all will undergo fertilization and proper development. Even if a healthy embryo is successfully tested, PGD does not work every time. Then only a fraction of those embryos that are successfully tested will have the "right" makeup. And IVF is a hit-and-miss affair, and more often than not the implantation of the embryo is unsuccessful. Success rests in a fraction of a fraction of a fraction of the fraction of eggs that are fertilized. There is no way to guarantee that even one IVF embryo can both be free of the damaged gene and have the right tissue type and successfully develop in the womb. There is a good chance that no child will result.

CLONING HEALTHY EMBRYOS

IMAGINE THAT YOU ARE the doctor who had to face a couple that has tried to have an unaffected child, in vain. IVF is grueling and distressing for couples who fail again and again to conceive. The emotional and financial cost of unsuccessful assisted conception can be high—so high that it may be better to give up, despite the heartbreak of childlessness. The creation of IVF embryos is easy enough. It is just that the overall success rate in getting an IVF pregnancy to term is relatively low, and we lack the techniques to correct genetic damage to improve that rate. Or rather, we lack a proven way to correct genetic damage accurately, confirm that it has worked, and only then transfer a modified human embryo to the womb.

In the medium term nuclear transfer does offer a way to do this. Combined with genetic manipulation, it allows precise DNA changes, including modification of existing genes. And it is possible to check that a genetic change has worked in the laboratory before a pregnancy is started. Since the birth of Polly these methods have been used to create entirely new biomedical opportunities, as was discussed in chapter 4. New methods for making genetic change are more efficient and accurate than that used so far. Precise change can be introduced without the need for selection by use of resistance to antibiotics.

The successful birth of Polly rested on the ability to grow cells in the laboratory for a long period during which genetic changes can be made. We used fetal cells to make her because they were available in sufficiently large numbers and have a long life in culture. However, the procedure of genetic change, selection of those cells with the modification, and confirmation of the change in our experience sometimes took too long. As a result the cells died or became abnormal in culture. This won't do for human medicine. For human medicine I envisage the use of embryonic stem cells, which can be grown almost indefinitely in culture, so that all the steps can be carried out. To prevent the birth of children with genetic disease, one would take an IVF embryo shown by preimplantation diagnosis to have a hereditary defect, remove the stem cells, carry out the genetic correction, check that modification has worked, then use nuclear transfer into an enucleated egg to create a new embryo without the disease.

The resulting child would be a clone not of any person but of an early embryo. To me this distinction is profoundly important. As I explained before, I am extremely concerned about the effects on a child of being a clone of another person, and I oppose it. However, an early embryo is not a person, and I see far fewer issues in the proposed use of nuclear transfer in this way to prevent a child's having a dreadful disease. One could even argue that this is not a true clone at all: although the genetic modification is slight, so that the cloned embryo's DNA differs by a fraction of a percent from that of the early embryo that donated the nucleus, the resulting child would be spared an awful hereditary disease. To a layperson there would be a world of difference between these clones.

No doubt there would be intense media interest in the first birth by this method, and that could be stressful for all those involved in the pregnancy. But in my view this stress would be less than that of being a "clone"—in the sense of being a genetically identical twin of another person—and would be imposed only on a few early cases, before the technique became routine and was deemed not newsworthy by journalists. If it were safe and effective, I can see nothing immoral about this use of nuclear transfer to prevent disease.

These methods could be developed to introduce genetic changes into embryos in order to prevent the birth of children with cystic fibrosis, who die prematurely of lung disease, or Huntington's disease, a devastating degenerative illness in which persons lose control of their limbs to such an extent that they appear to dance. The original name for the disease, Huntington's chorea, means just that— "Huntington's dance" ("chorea" comes from the Greek word for dancing). The first symptoms do not occur until middle age but the decline is relentless, ending in death. Not surprisingly, young people at risk of this awful disease often reject the offer of a genetic test to show whether it will develop.

My proposal would not cure afflicted parents but would help them ensure that they have children free of the disease. If an affected couple produced an embryo that carried the abnormal gene, revealed by a telltale "genetic stutter" in the genetic code typical of Huntington's, then genetic modification of those cells before cloning to produce a new, stutter-free embryo would result in a child without the disease. When it becomes safe to do so, I strongly believe, we should consider using cloning for this purpose. This method offers new hope to those who are unlucky and who consistently fail to produce healthy embryos. When I ask students what they think of using genetic correction and cloning this way, they point out the difficulty of doing it, the scarcity of eggs, and so on. But they raise no fundamental ethical objections. I suspect—and I hope—the public will feel the same way.

ENHANCING THE
GERM LINE

THE CHILD PRODUCED as a result of nuclear transfer using a genetically altered embryonic cell would grow to be an adult who could pass on this genetic change. This is germ line modification. A fertility treatment called cytoplasmic transfer has some of the same characteristics. In this experimental method the cytoplasm from a healthy, donor egg is injected into the egg of an older woman to "rejuvenate" it. The purpose of cytoplasmic transfer is ultimately to produce an embryo with the genetic makeup of both parents, providing an alternative to using donated eggs. But it may also have the effect of introducing a genetic change—foreign DNA in the mitochondrial power packs in the donor cytoplasm—that may be passed to future generations down the maternal line (we inherit only our mother's power packs). Since the first child, Emma, was born in 1997, a small numbers of babies have been created this way, and, in a limited respect, they have two mothers.

With regard to efficacy a number of prominent reproductive biologists, including Lord Winston, Alan Trounson, and Roger Short were unconvinced that cytoplasmic transfer improved pregnancy rates. And although mitochondria are usually thought of as governing cellular energy production, rather than the phenotype (the outward physical appearance and constitution of a creature, influenced by DNA in the nucleus), there were concerns about unforeseen problems because they also play important roles in other cellular processes (such as cell death). The controversial work, which was stopped in 2001, was done at the Institute for Reproductive Medicine and Science of St. Barnabas in Livingston, New Jersey, by a team that included a familiar figure in the cloning story, none other than Steen Willadsen.

The therapy I envisage would go much further and alter one in thousands of the instructions in the nuclear DNA, the DNA that shapes appearance, intellect, and susceptibility to disease. Many scientists are convinced that we should never attempt this because the consequences are passed to future generations. But not everyone disapproves. "Whether based in ethics or unfounded fears of the unknown, such arguments are ultimately not compelling in my judgment," asserts Jim Watson, who helped launch the molecular biology revolution with Francis Crick and Maurice Wilkins by revealing the structure of DNA. "Germ line therapy is in principle simply putting right what chance has put horribly wrong." This is a view that I share.

In his book *DNA: The Secret of Life*, Watson goes beyond the relatively uncontroversial application of fixing hereditary disease, to preventing disease in the first place:

> Germ line gene therapy has the potential for making humans resistant to the ravages of HIV. Some would say that, rather than altering people's genes, we should concentrate our efforts on treating those we can and impressing upon everyone else the dangers of promiscuous sex. But I find such a moralistic response to be profoundly *immoral*. Education has proven a powerful but hopelessly insufficient weapon in our war.

Indeed, Watson goes further still, advocating genetic enhancement: genetic changes that address our desires rather than needs. This could lead to a social revolution, since we would be able—in theory, at least—to mold our children in all kinds of ways. In raising the possibility of viable germ line gene therapy, he says, "Having identified the relevant genes, would we want to exercise a future power to transform slow learners into fast ones before they are even born? We are not dealing in science fiction here: we can already give mice better memories. Is there a reason why our goal shouldn't be to do the same for humans?" Despite the risks, Watson argues that we should give serious consideration to germ line gene therapy, adding,

> I only hope that the many biologists who share my opinion will stand tall in the debates to come and not be intimidated by the inevitable criticism. Some of us already know the pain of being tarred with the brush once reserved for eugenicists. But that is ultimately a small price to pay to redress genetic injustice. If such work be called eugenics, then I am a eugenicist.

One could argue that the blend of cloning and genetic modification can offer another route to enhancement, so that parents can offer their children a "better life." In one sense there is nothing new in parents' struggling to give their children a head start. They try to get them into the best schools. They may sign them up for additional classes, whether in pottery, music, or ballet. They ensure they are vaccinated. They give them antibiotics and warm them with central heating to keep them well. They pay for expensive dental work to correct their teeth. Some even sign up their children for cosmetic surgery, or alter their mood with drugs such as Ritalin and Prozac.

I suspect that many people would consider tinkering with the genes of their children too. Proponents of enhancement would accept that genes do not determine precisely who we are but argue that they constrain an envelope of possibilities, so genetic alteration offers a way to increase the range of possibilities. For example, it might be used to alter genes of utility, such as those linked with memory or dexterity. It might alter genes valued by society, such as those that contribute to our capacity for altruism and sympathy. There are even wilder ideas, of altering humans so that they are more tolerant of pollution, or of climate change.

I have objections to the proposal to enhance the abilities of our children on both biological and ethical grounds. Like most people, I disapprove strongly of the idea of an embryo coaxed to life for shallow reasons of status, preference, or style. Any such work is unsavory because it reduces children to consumer objects that can be "accessorized" according to the parents' whims. As many ethicists have argued, love for offspring should not be contingent upon the characteristics they possess, even in our less than ideal world.

There is also a fundamental difference between genetic enhancement and established methods of enhancement, such as education, mood-altering drugs, and cosmetic surgery. Genetic changes would be transmitted to all future generations. And their effects would be harder to reverse. They could undermine qualities and traits that are fundamental to our humanness. They could exaggerate intolerance of disabilities. There is a need for great caution.

There is also a major biological concern, since I am skeptical that genetic enhancement is even possible because the genetic control of many traits is so complex. In turn any of the genes involved in any one trait also has effects in many others. In short, we could not predict the effect of changing genes except in the case of inherited diseases associated with errors in a known gene. Discussions of designer babies often refer to how genes control eye color, which they do. And so follow many glib claims about creating blue-eyed blond-haired babies. But even in this simple example the genetics is far from straightforward. The color of our eyes is influenced by the interaction of several genes that alter the distribution and content of pigment (melanin) producing cells, melanocytes, in the eye. Children are usually taught that brown is dominant, so if one parent has brown eyes their offspring can also have them, while two blue-eyed parents always have a blue-eyed child. But, though uncommon, blue-eyed parents can have children with brown eyes. Indeed, there are subtleties of eye color and a range of hues that scientists cannot yet fully explain.

With respect to such traits as intelligence, in addition to this genetic complexity, the environment—our family life, our education—has of course a considerable effect, and there is a limit to what can be achieved by snipping, inserting, and shuffling DNA. Like most diseases, most traits will turn out to be polygenetic—the product of an interplay of genes with one another and with the environment. Intelligence, facial structure, and extroversion are likely to be genetically complex. Even if it were possible to spot the crucial genes, enormous numbers of embryos would have to be tested to find the one with "the right stuff."

There is another problem facing parents who want to boost the brainpower of their child: genes can influence several characteristics. Any genetic change intended to influence intelligence, say, could also change other aspects of personality in an unpredictable way. One of the human genome pioneers, Francis Collins, told a major conference in France a few years ago about the rich parents who hoped to produce a baby genetically enhanced to be artistic and musical but ended up with "a sullen adolescent who smokes marijuana and doesn't talk to them." In short, each "enhancement" would be an experiment. I find it unacceptable to experiment on children. I believe that we ought to accept ourselves and our children for what we are, rather than attempting to breed an improved race of human beings. Like all my peers, I think that money would be better spent on education than on genetic enhancement. Finding a good school will be the best way to enhance a child's intelligence for some considerable time.

Others have extended conjecture about the genetic engineering to lengths that have no bearing upon reality. Lee M. Silver of Princeton University, the author of *Remaking Eden: Cloning and Beyond in a Brave New World*, raises the specter of boutique babies in which parents sit at a computer and scroll through a series of genetic menus as they choose the characteristics they want for their offspring. He speculates how, in the distant future, the upper classes will have had turbocharged children with so much genetic enhancement over so many generations that they become unable to mate with members of other classes. The posh would be a different species to paupers, and he warns that this would be "the most horrible thing that ever happened to humanity."

I have confidence that posh and paupers will continue to exchange genes, whether by old-fashioned sex or by assisted reproduction, so that divergence into separate species will not occur. I appreciate that Silver is trying to make a vivid point, but this speculation would be more at home in a science-fiction novel. We have been here before, after all, with the beautiful Eloi and the brutish Morlocks of *The Time Machine*, the 1895 classic by H. G. Wells. However, I do accept Silver's point that the driving force behind many uses of new technology is not the government, as in Aldous Huxley's *Brave New World*, with its state-run human hatcheries. Rather, it is commercial pressures. If we leave it to market forces, then I am sure some couples will attempt to design their own children. If fictions like *Remaking Eden* or *Brave New World* teach us anything, it is that the unfettered use of scientific technologies can present a danger to human freedom and dignity. The way that they are applied has to be regulated. As Huxley once said, science in itself is morally neutral; it becomes good or evil according to how it is applied.

BANNING DESIGNER BABIES

DESPITE MANY PEOPLE'S almost religious belief in free markets, our experience of the food and pharmaceutical industries shows us that an unregulated market is inappropriate for matters of public health and thus of genetic modification. In the United States, where the faith in market forces is strong, the production of food and medicines has been supervised for a century by the Food and Drug Administration.

At the beginning of the twentieth century there was little regulation of the sale of medicines. Many of the products that were sold had no clinical value. Those that did, like quinine-containing cinchona bark powder, could be made less effective—and much more profitable—by cutting them with just about anything, and others were laced with what we now recognize as recreational drugs. The latter may have reduced discomfort and perhaps made the patient "happy" for a few hours, but they provided no treatment of the underlying condition.

Supervision of food production and marketing was also very limited. Upton Sinclair's influential novel *The Jungle* (1906) described the wretched sanitary and working conditions in the Chicago meatpacking industry and helped bring about action. Publications such as the *Ladies' Home Journal* played a leading role in bringing political pressure to bear in Congress. A number of acts to tighten the regulation of production and marketing were passed. Despite these changes there was still no obligation to assess the effect of new medicines, and tragedy was inevitable. A Tennessee drug company marketed a form of the new wonder drug Elixir Sulfanilamide, an early antibiotic. However, the sweet-tasting solvent in this untested product was a highly toxic chemical analogue of antifreeze; over a hundred people died, including many children. The public outcry led to Franklin D. Roosevelt's signing the Food, Drug, and Cosmetic Act on 25 June 1938. The thalidomide tragedy of the 1960s further expanded the FDA's authority. The moral of these grim stories is clear: it often takes a disaster to force governments to take action.

Past events warn us of the very real danger of tragic consequences if developments in reproductive technology escape regulation. In particular, great care is needed if the entire future of a child is being determined. A regulatory framework is required. This is too important to leave to the scientists, the clinicians, or even the prospective parents, each of whom has their own specific viewpoint. Rather, we should expect a framework to be prepared to guide those who are too closely involved, who may have vested interests, and who may fail to appreciate the full consequences for the children who result. The framework could be enforced by a national or regional regulation, or by a hospital ethics committee. The body that provides the framework is not really important. The critical point is that guidance be provided by someone who is not on the cast list of those who are directly involved.

DRAWING THE LINE

ALTHOUGH I AM for the use of genetic modification to treat disease and against enhancement, I would be the first to admit that there will be endless arguments over where to draw the line. Everyone understands the wish to keep children from being born with a serious genetic disease such as Huntington's, for which there is no treatment and whose consequences are devastating. If it becomes safe and effective to prevent this and other grave genetic disorders by use of cloning, it is hard to see how this will be ethically objectionable. Similarly, most of us would judge it desirable to administer synthetic growth hormone to ensure that a stunted child grows to normal height. By the same token, the use of germ line therapy to spare the children of that child a similar fate seems reasonable. But what if a "normal" child's parents cited the extensive and persuasive scientific literature on how tall people fare better in life—they tend to earn more, in terms of respect, partners, and money—and demand the height of their normal child be boosted a little more? Indeed, what if the parents wanted to guarantee that their child would be tall enough to be a basketball player?

An inadequate, abnormal, or unusual physical appearance can lead to depression and feelings of inadequacy. As a consequence, some parents may argue for cosmetic changes that they believe could even save a life by preventing vulnerable young persons from becoming suicidal about the size of their nose, their baldness, and so on. The risks that people take with cosmetic surgery are a powerful testament to the desire for an acceptable body image, and there will doubtless be intense pressure from some parents to carry out genetic surgery on their unborn child to achieve similar ends.

The possibilities of genetic tinkering raise other issues. Is aging a disease? Surely not; it is a normal part of the life cycle. But, given recent research that has extended the life of creatures such as the nematode worm, it seems highly likely that there are human genes that influence longevity (though they are by no means the whole story when it comes to a long life). Does that mean old age should be "cured" with genetic modification? And if so, what does that mean for society? John Harris, a professor of bioethics at Manchester University, has suggested this "creeping longevity" would have profound implications for what we mean by the sanctity of human life, increasing resentment as the super-old competed with the young for jobs, space, and other resources, may even lead to what he calls "generational cleansing."

Whatever the shade of gray between enhancement and therapy and whatever boundary is being transgressed, one aspect of this medical intervention stays the same: genetic enhancement by the method I have outlined in this chapter will be proposed by parents on behalf of their unborn child. This is about the fate of children of the future, a decision that only has indirect impacts on the person making the choice. Society has an obligation to intervene on the embryo's behalf when it comes to weighing risks and benefits of genetic alteration. Selection for traits thought by the parents to be beneficial could be seen as a curse by the child, as parents bear down to achieve their goals and to make their investment worthwhile. Uninhibited selection of children may erode the unconditional love that is the bedrock of the parent-child relationship.

Aside from these fundamental moral and ethical issues, there are practical and social considerations. Genetic correction will be expensive. Several complex and difficult steps are required, and at each point careful checks must be made to ensure that all is well. To make this process reliable will take time and cost money. This means that choices will have to be made about how to allocate finite resources. As ever, the benefits are most likely to be enjoyed by the rich first, suggesting there will be genetic haves and have-nots. Hospitals and health providers are already well versed in this issue. The use of this method could also be curbed by the limited availability of eggs, which have other uses, notably in the treatment of infertility. In short, the genetic modification of the human gene pool will have to be considered in a broad context, that of society as well as the family. And perhaps that of history too.

In many of the discussions of where my research is taking us, one senses that the benchmark used to judge whether reproductive science is helpful or harmful is a utopian zero-risk society, a society free of narcissism, a society where children are much more than chattels, and a just society free of the taint of self-interest. But over the millennia, pressure from family, peers, and society have thrown together couples to bind allegiances, achieve status, and to combine wealth. Children have been conceived to perpetuate businesses, estates, and royal bloodlines. In a sense, designer babies are nothing new: think of the mother with three daughters who is desperate to conceive again to have a son; the career woman approaching menopause who feels she had better conceive quickly so that she can claim to have led a Full Life; the poverty-stricken who have a big family to help work the farm; and the celebrity who thinks that a baby in designer denim is this year's must-have accessory.

The effects of these traditional splashings in the human gene pool, not utopia, should provide the upper benchmark for what should and should not be allowed when we apply reproductive science: we should aspire to do much better than this and hope that we never do any worse. Whatever we do, I doubt that this science will cause more than ripples when compared with the waves of evolutionary change that came before.

NINE

BEYOND
HUMAN CLONING

THE SPIN, hype, and exaggeration that have so far greeted my work suggest a dizzying pace of progress in research on assisted reproduction carried out by wild-eyed Frankensteins, with science out of control and accelerating far too quickly for irritating little things like ethical considerations to have the smallest hope of keeping up.

The truth, instead, is that we have plenty of time to sort out the ethics, because relative to scientific progress, the application of this knowledge in reproductive technology is slow, painstaking, and already highly regulated in many countries. Take the development of IVF, for example. Beginning in earnest in 1962, IVF research achieved many technical milestones, from egg maturation and fertilization in the test tube to methods to stimulate the growth of follicles and so on. As if to underline the revolutionary implications of what they were trying to do, Robert Edwards and Patrick Steptoe were denied government funds and attacked by the "Holy Trinity": the pope, the press, and prominent Nobel laureates (Jim Watson demanded of Edwards at a 1971 press conference, "What are we going to do with the mistakes?").

IVF embryo transfers began only in 1972, a decade after IVF research began, and were rewarded by a first clinical pregnancy in 1976, unfortunately ectopic. The effort finally culminated in the birth of Louise Brown in 1978. The technique has since been shown to be as safe as one could expect and has brought joy to millions of people around the world. Debate continues, of course. Since the birth of Louise Brown there have been arguments about whether IVF should be used to enable postmenopausal women to have children, what the fate of spare and frozen embryos should be, and whether every woman has the right to have a child. But we have had many years to mull the implications over. This long debate and the obvious benefits for the infertile have built public confidence in IVF.

A similar history of early failure and social controversy followed by eventual technical success and public acceptance could be charted for many biological innovations, including organ transplantation and the introduction of anesthetic during childbirth. Countries differ in the areas in which they will be ambitious and the extent to which they are adventurous, but the general pattern is similar everywhere.

We are also fortunate in that researchers have led many discussions of the implications of their work. I hope with this book to continue a long tradition in which scientists are open about what they do and in speculating about where their research is taking us. In the case of germ line modification, a paper published in the journal *Science* in 1972 shows that scientists were thinking of the consequences even then. And they continue to do so, for instance, in a recent report by the American Association for the Advancement of Science. Moreover, we have active media that show a great, sometimes overactive, imagination when it comes to revealing the future path and pitfalls of a particular technology.

But I have faith in more than openness, debate, and regulation. I have faith in people too. I have confidence that a well-informed democracy can keep abuses in check; I have confidence in women who would not donate eggs to clone a dictator, but would to help a patient with an urgent clinical need; above all else I have faith in the vast majority of scientists, who are no different from anyone else in wanting to reduce suffering and make the world a better place.

There is one crucial caveat I must add. No matter how powerful science becomes, the future will remain mysterious and open. You can predict neither the outcome of research nor the uses that will be made of new knowledge. The birth of Dolly opened the door onto a new realm of medicine and biology. After Dolly we can glimpse some of the treasures and pitfalls in this realm, but we still have much more to learn.

Just as the influence of the Internet today could never have been anticipated when Tim Berners-Lee invented the World Wide Web in 1989, so we should also expect that methods of cloning will be unrecognizable twenty years from now. As for what will be possible in a century, that is anybody's guess, though I imagine that someone out there, whether scientist, futurist, or science-fiction writer, has already thought about it to some extent.

Whatever emerges, I am confident that the technology will become safer, that there will be embryo-free alternatives to the use of nuclear transfer in the creation of stem cells, and that its combination with genetic modification will become routine. Most important, I am sure that endless debate will thrash out many of the difficult ethical issues to the satisfaction of the majority of society.

It is critical that we do not allow fear of misuse of new knowledge to curb our exuberant creativity. We often take for granted what has been hard-won from previous research, and overlook the possibilities and potential of future developments that may well dwarf the attainments of the past. As much damage can be done by failing to exploit the beneficial applications of a technology as by promoting the applications of that technology which are risky or harmful.

Assisted reproduction technology is very much in its stone age, and the revulsion that my proposal will stir in some people says less about the technology itself and more about the shock of the new, as it once did in the early days of heart transplants, vaccines, and other innovations. The future-that-was that came in the wake of these early developments shows how utility can breed acceptance.

The case of genetic alteration of humans I have outlined in this book will always have its critics. Even when the technologies of nuclear transfer, genetic manipulation, and stem cells have matured, I am sure that some people will still prefer to put up with the random insults of nature than be subject to human intervention, even if it is based on careful consideration of medical issues rather than a whim. They are, of course, free to turn their backs on the benefits of new technology. But at least they will have a choice. And for me, just having the chance to decide is paramount.

Although I have no idea what the future will bring, I am confident that reproductive and genetic technologies will greatly expand our possibilities. By the same token, they will also expand the burden of responsibility. Because of my faith in the majority of people to know right from wrong, I feel the sooner we take on that burden the better.

I want to be able to change my destiny rather than be condemned to a particular fate. I want people to have new options when it comes to that most fundamental urge to bring healthy children into this world. For me the widespread use of genetic and reproductive technologies is not a step backwards into darkness but a step forwards into the light.

SOURCE NOTES

INTRODUCTION

Interviews, Correspondences, and Speeches

Campbell, Keith. E-mail correspondence with Roger Highfield, 4 and 5 July 2005.

———. Interview with Roger Highfield, 16 June 2005.

Harkin, Senator Thomas. "Human Cloning Will Take Place." Testimony to the Labor and Human Resources Committee, 12 March 1997.

Kendrick, Keith. E-mail correspondence with Roger Highfield, 23 Aug. 2004 and 26 April and 12 Aug. 2005.

Kennedy, Senator Edward M. "Sound Barrier." Testimony to the Labor and Human Resources Committee, 12 March 1997.

King, Tim. E-mail correspondence with Roger Highfield, 6 April and 16 June 2005.

———. Interview with Roger Highfield, Roslin Institute, 14 June 2005.

Miller, Dusty. E-mail correspondence with Roger Highfield, 11 and 14 April 2005.

Rhind, Susan. Phone interview with Roger Highfield, 28 March 2005.

Weldon, Fay. Conversations with Roger Highfield, Berkeley Hotel, London, 3 Nov. 1994; Savoy, London, 21 Nov. 2002 and 24 Feb. 2004; Groucho Club, London, 26 Sept. 2003. Joint visit to the Roslin Institute, 18 Nov. 2003.

Wilmut, Ian. Testimony to the Labor and Human Resources Committee, 12 March 1997.

Zavos, Panos. Press conference, London, 26 Sept. 2005.

Journal and Newspaper Articles

Cohen, Philip. "Into the Clone Zone." *New Scientist* 151, no. 2133 (9 May 1998): 3232.

Cox News Service. "Unexpected Appearance of Scotland's Cloned Sheep." 25 June 1997.

Dyer, Geoff, and David Firn. "Death of Dolly the Sheep Fuels Fears about Cloning." *Financial Times*, 15 Feb. 2003, p. 3.

Grice, Elizabeth. "My Long Search for Liberty." *Daily Telegraph*, 19 Jan. 1998, p. 13.

Highfield, Roger. "British Scientists Condemn Plans to Clone Babies." *Daily Telegraph*, 10 March 2001, p. 8.

———. "Fertility Experts Expel Human Clone Doctor." *Daily Telegraph*, 15 Sept. 2001, p. 18.

———. "From Science Fiction to Unproved Fact." *Daily Telegraph*, 12 Aug. 2004, p. 4.

———. "Human Clone Claim Is Met with Scepticism." *Daily Telegraph*, 28 Nov. 2002, p. 9.

———. "Rival Joins Race to Clone First Human." *Daily Telegraph*, 3 Dec. 2002, p. 8.

———. "Scientists Boycott 'Human Clone' Conference." *Daily Telegraph*, 12 Sept. 2001, p. 11.

———. "US Fertility Doctor Reveals Another Human Cloning Failure." *Daily Telegraph*, 27 Sept. 2005, p. 2

———. "We've Cloned Baby, Says Cult." *Daily Telegraph*, 29 Dec. 2002, p. 1.

Jaenisch, Rudolf, and Ian Wilmut. "Don't Clone Humans!" *Science* 291 (2001): 2552.

Kolata, Gina. "Some Scientists Ask: How Do We Know Dolly Is a Clone?" *New York Times*, 29 July 1997, p. C3.

Mestel, Rosie. "Dolly's Demise Puts Cloning into Question." *Orlando Sentinel*, 23 Feb. 2003, p. G6.

Milne, Shaun. "Dolly Makes Opera Debut at Baa-rbican." *Mirror*, 17 Sept. 2002, p. 10.

Pennisi, Elizabeth. "Cloning: The Lamb That Roared." *Science* 278 (1997): 2038–39.

Reuters. "Scientists Clone the Unexpected." *Chicago Tribune*, 25 Feb. 1997, p. 1.

Vendantam, Shankar. "British Scientist Who Cloned Sheep Urges Caution at Senate Hearing." Knight Ridder/Tribune News Service, 12 March 1997.

Wade, Nicholas. "Scientist Who Announced Cloning of Sheep Dolly Is Taking Steps to Win Over Critics." *New York Times*, 28 Feb. 1998, p. A6.

Watson, J. D., and F. H. C. Crick. "A Structure for Deoxyribose Nucleic Acid." *Nature* 171 (1953): 737–38.

Wooton, Sarah K., Christine L. Halbert, and A. Dusty Miller. "Sheep Retrovirus Structural Protein Induces Lung Tumors." *Nature* 434 (14 April 2005): 904–7.

Books

Churchill, Caryl. *A Number*. London: Nick Hern, 2002.

Kolata, Gina. *Clone: The Road to Dolly, and the Path Ahead*. New York: Morrow, 1998.

Watson, J. D. *The Double Helix: A Personal Account of the Discovery of the Structure of DNA*. Edited by Gunther Stent. New York: Norton, 1980.

Wilmut, Ian, Keith Campbell, and Colin Tudge. *The Second Creation: Dolly and the Age of Biological Control*. New York: Farrar, Straus and Giroux, 2000.

Web Sites

National Right to Life Homepage. http://www.nrlc.org.

Raelian Movement Homepage. http://www.rael.org.

Reich, Steve, and Beryl Korot. "Three Tales." Steven Reich Homepage .http://www.stevereich.com.

Siegel, Bernard. Genetics Policy Institute Homepage. http://www .genpol.org.

Zavos Organization Homepage. http://www.zavos.org.

CHAPTER 1: CLONING THE CLONER

Interviews, Correspondences, and Speeches

Evans, Martin. Interview with Roger Highfield, Biochemical Society meeting, Nottingham, 20 July 1988, and phone interview with Roger Highfield, 9 March 2005.

Polge, Chris. Interview with Roger Highfield, Royal Society, London, 23 June 1987.

Journal and Newspaper Articles

Aréchaga, J. "Embryo Culture, Stem Cells and Experimental Modification of Embryonic Genoma: An Interview with Ralph Brinster." *International Journal of Developmental Biology* 42 (1998): 861–77.

Brinster, R. L. "Germline Stem Cell Transplantation and Transgenesis." *Science* 296 (2002): 2174–76.

Emlen, S. T. "The Evolutionary Study of Human Family Systems." *Social Science Information* 36 (1997): 563–89.

Evans, M. J., and M. H. Kaufman. "Establishment in Culture of Pluripotent Cells from Mouse Embryos." *Nature* 292 (1981): 154–56.

Gordon, J. W., G. A. Scangos, D. J. Plotkin, J. A. Barbosa, and F. H. Ruddle. "Genetic Transformation of Mouse Embryos by Microinjection of Purified DNA." *Proceedings of the National Academy of Sciences* 77 (1980): 7380–84.

Hammer, R. E., V. G. Pursel, C. E. Rexroad Jr., et al. "Production of Transgenic Rabbits, Sheep and Pigs by Microinjection." *Nature* 315 (1985): 680–83.

Highfield, Roger. "Brave New Era Dawning Down on the Farm." *Daily Telegraph*, 24 June 1987, p. 4.

———. "Chimeras Help Scientists in Battle against Genetic Illness." *Daily Telegraph*, 21 July 1988, p. 8.

Polge, C., S. Salamon, and Ian Wilmut. "Fertilizing Capacity of Frozen Boar Semen following Surgical Insemination." *Veterinary Record* 87 (1970): 424.

Simons, J. P., I. Wilmut, I., A. J. Clark, A. L. Archibald, J. O. Bishop, and R. Lathe. "Gene Transfer into Sheep." *Bio/Technology* 6 (1988): 179.

Wilmut, I., and L. Rowson. "Successful Low-Temperature Preservation of Mouse and Cow Embryos." *Journal of Reproduction and Fertility* 33 (1973): 352–53.

———. "Experiments on Low-Temperature Preservation of Cow Embryos." *Veterinary Record* 92 (1973): 686–90.

Books

Mori/Department of Trade and Industry. *Science in Society: Findings from Qualitative and Quantitative Research*. March 2005.

Levick, Stephen E. *Clone Being: Exploring the Psychological and Social Dimensions*. Lanham, Md.: Rowman and Littlefield, 2004.

Web Sites

Department of Trade and Industry Homepage. http://www.dti.gov.uk.

Roslin Institute: Edinburgh Homepage. http://www.roslin.ac.uk.

Rosslyn Chapel Homepage. http://www.rosslynchapel.org.uk.

CHAPTER 2: A BRIEF HISTORY OF CLONING

Interviews, Correspondences, and Speeches

Bromhall, Derek. E-mail correspondence with Roger Highfield, 26 April and 11 May 2005.

Emlen, Stephen T. "Sexual Differences in Mating Strategies: The Interface of Culture, Law and Biology." American Association for the Advancement of Science, Seattle, 14 Feb. 1997.

Galli, Cesare. E-mail correspondence with Roger Highfield, 13 and 14 April 2003.

Graham, Chris. E-mail correspondence with Roger Highfield, 15 April 2005.

———. Phone interview with Roger Highfield, 15 April 2005.

Gurdon, John. E-mail correspondence with Roger Highfield, 14 and 22 March, 1 and 26 April, and 12 July 2005.

Ritchie, Marjorie. Interview with Roger Highfield, Roslin Institute, 14 June 2005.

Schatten, Gerald. E-mail correspondence with Roger Highfield, 27 Jan., 6 March, and 1 Aug. 2005.

———. Interview by phone with Roger Highfield, 17 May 2005.

Solter, Davor. E-mail correspondence with Roger Highfield, 16, 18, and 30 March, 4, 7, and 19 April, 31 May, and 4 June 2005.

Tarkowski, Andrzej. E-mail correspondence with Roger Highfield, 25 and 26 April and 6 May 2005.

Willadsen, Steen. E-mail correspondence with Roger Highfield, 12 Oct. 2004.

Journal and Newspaper Articles

Briggs, R., and T. J. King. "Nuclear Transplantation Studies on the Early Gastrula (*Rana pipiens*): I. Nuclei of Presumptive Endoderm." *Developmental Biology* 2 (1960): 252–70.

———. "Transplantation of Living Nuclei from Blastula Cells into Enucleated Frogs' Eggs." *Proceedings of the National Academy of Sciences* 38 (1952): 455–63.

Bromhall, J. D. "Nuclear Transplantation in the Rabbit Egg." *Nature* 258 (1975): 719–22.

Budiansky, S. "Karl Illmensee: NIH Withdraws Research Grant." *Nature* 309 (1984): 738.

Campbell, K. H. S., J. McWhir, W. A. Ritchie, and I. Wilmut. "Sheep Cloned by Nuclear Transfer from a Cultured Cell Line." *Nature* 380 (1996): 64–66.

Chan, A. W. S., T. Dominko, C. M. Luetjens, et al. "Clonal Propagation of Primate Offspring by Embryo Splitting." *Science* 287 (2000): 317–19.

Driesch, Hans. "Zur Verlagerung der Blastomeren des Echinideneies." *Anatomischer Anzeiger* 8 (1893): 348–57.

Fishberg, M., J. B. Gurdon, and T. R. Elsdale. "Sexually Mature Individuals of *Xenopus laevis* from the Transplantation of Single Somatic Nuclei." *Nature* 182 (1958): 64–65.

Gurdon, J. B., and J. A. Byrne. "The First Half-century of Nuclear Transplantation." *Proceedings of the National Academy of Sciences* 100 (2003): 8048–52.

Highfield, Roger. "Cloned Sheep Have Endless Ramifications—Where Sheep Are Sheep and Men Are Uneasy." *Daily Telegraph*, 7 March 1996, p. 1.

———. "Designer Sperm 'Can Weed Out' Gene Defects." *Daily Telegraph*, 23 Oct. 2001, p. 13.

Hoppe, P. C., and K. Illmensee. "Microsurgically Produced Homozygous-Diploid Uniparental Mice." *Proceedings of the National Academy of Sciences* 74, 12 (1977): 5657–61.

Illmensee, K., and P. C. Hoppe. "Nuclear Transplantation in *Mus musculus*: Developmental Potential of Nuclei from Preimplantation Embryos." *Cell* 23 (1981): 9–18.

Marx, J. L. "Bar Harbor Investigation Reveals No Fraud." *Science* 220 (1983): 1254.

McGrath, J., and D. Solter. "Inability of Mouse Blastomere Nuclei Transferred to Enucleated Zygotes to Support Development in Vitro." *Science* 226 (1984): 1317–19.

———. "Nuclear Transplantation in the Mouse Embryo by Microsurgery and Cell Fusion." *Science* 220 (1983): 1300.

McLaren, A. "Cloning: Pathways to a Pluripotent Future." *Science* 288 (2000): 1775–80.

Newmark, P. "Geneva and Bar Harbor Labs to Check Results." *Nature* 303 (1983): 363.

Solter, D. "Imprinting." *International Journal of Developmental Biology* 42 (1998): 951–54.

———. "Mammalian Cloning: Advances and Limitations." *Nature Reviews: Genetics* 1 no. 3 (2000): 199–207.

Tarkowski, A. K. "Mouse Chimaeras Revisited: Recollections and Reflections." *International Journal of Developmental Biology* 42 (1998): 903–8.

Tsunoda, Y., and Y. Kato. "Full-Term Development after Transfer of Nuclei from 4-Cell and Compacted Morula Stage Embryos to Enucleated Oocytes in the Mouse." *Journal of Experimental Zoology* 278 (1997): 250–54.

Wakayama, T., I. Rodriguez, A. C. F. Perry, R. Yanagimachi, and P. Mombaerts. "Mice Cloned from Embryonic Stem Cells." *Proceedings of the National Academy of Sciences* 96 (1999): 14984–89.

Willadsen, S. M. "Nuclear Transplantation in Sheep Embryos." *Nature* 320 (1986): 63–65.

Woods, G. L., K. L. White, D. K. Vanderwall, et al. "A Mule Cloned from Fetal Cells by Nuclear Transfer." *Science* 301 (2003): 1063.

Books

Cibelli, J., R. P. Lanza, K. H. S. Campbell, and M. D. West, eds. *Principles of Cloning*. San Diego: Academic Press, 2002.

Huxley, Aldous. *Brave New World*. New York: Harper, 1998.

Mitchison, Naomi. *Solution Three*. New York: Feminist Press, 1995.

Rorvik, David. *In His Image: The Cloning of a Man*. Philadelphia: Lippincott, 1978.

Spemann, Hans. *Embryonic Development and Induction*. New Haven: Yale University Press, 1938.

Toffler, Alvin. *Future Shock*. New York: Random House, 1970.

Weismann, August. *The Germ-Plasm: A Theory of Heredity*. New York: Scribner's, 1893.

Wilmut, Ian, Keith Campbell, and Colin Tudge. *The Second Creation: Dolly and the Age of Biological Control*. New York: Farrar, Straus and Giroux, 2000.

Web Sites

Brownlee, Christen. "Nuclear Transfer: Bringing in the Clones." Proceedings of the National Academy of Sciences. http://www.pnas.org/misc/classics4.shtml.
Genetics Savings and Clone, Inc. Homepage. http://www.savingsandclone.com.
Nobel Prize Organization Homepage. http://www.nobelprize.org.

Unpublished Material

Brown, Louise. "The Baby Makers." London, Channel 4, 1999.

CHAPTER 3: DOLLYMANIA

Interviews, Correspondences, and Speeches

Archibald, Alan. E-mail correspondence with Roger Highfield, 10 Feb. 2005.
———. Interview with Roger Highfield, Roslin Institute, 18 Nov. 2003.
Bracken, John. Interview with Roger Highfield, Roslin Institute, 14 June 2005.
Campbell, Keith. E-mail correspondence with Roger Highfield, 4 and 5 July 2005.
———. Phone interview with Roger Highfield, 16 June 2005.
Colman, Alan. E-mail correspondence with Roger Highfield, 14 April 2005.
Larsson, Nils-Göran. E-mail correspondence with Roger Highfield, 26 May 2004.
Ritchie, Marjorie. Interview with Roger Highfield, Roslin Institute, 14 June 2005.

Walker, Karen (née Mycock). E-mail correspondence with Roger Highfield, 7, 8, 12, and 24 July 2005.

Yang, Jerry. E-mail correspondence with Roger Highfield, 14 April and 23 June 2005.

Journal and Newspaper Articles

Ashworth, D., M. Bishop, K. Campbell, et al. "DNA Microsatellite Analysis of Dolly." *Nature* 394 (1998): 329.

Campbell, K. H. S., P. Loi, P. J. Otaegui, and I. Wilmut. "Cell Cycle Co-ordination in Embryo Cloning by Nuclear Transfer." *Reviews of Reproduction* 1 (1996): 40–46.

Galli, C., I. Lagutina, G. Crotti, et al. "A Cloned Horse Born to Its Dam Twin." *Nature* 424 (2003): 635.

Highfield, Roger. "She Looks Exactly like Any Other Sheep. But the Cloning Method Used to Produce Dolly May Change Our Lives." *Daily Telegraph*, 24 Feb. 1997, p. 5.

Mckie, Robin. "Scientists Clone Adult Sheep." *Observer*, 23 Feb. 1997, p. 1.

Pell, Cardinal George. "Slippery Slope." *Sunday Telegraph*, 26 June 2005, p. 99.

Signer, E., Y. E. Dubrova, A. J. Jeffreys, et al. "DNA Fingerprinting Dolly." *Nature* 394 (1998): 329–30.

Solter, D. "Dolly Is a Clone—And No Longer Alone." *Nature* 394 (1998): 315–16.

———. "Lambing by Nuclear Transfer." *Nature* 380 (1996): 24.

Thompson J. G., D. K. Gardner, P. A. Pugh, W. H. McMillan, and H. R. Tervit. "Lamb Birth Weight Is Affected by Culture System Utilized during in Vitro Pre-elongation Development of Ovine Embryos." *Biology of Reproduction* 53 (1995): 1385–91.

Tian, X. C., C. Kubota, K. Sakashita, et al. "Meat and Milk Compositions of Bovine Clones." *Proceedings of the National Academy of Sciences* 102 (2005): 6261–66.

Vittorio, S., and N. D. Zinder. "Letters." *Science* 279 (1998): 635–36.

Wakayama, T., A. C. Perry, M. Zuccotti, K. R. Johnson, and R. Yanagimachi. "Full-term Development of Mice from Enucleated

Oocytes Injected with Cumulus Cell Nuclei." *Nature* 394 (1998): 369–74.

Walker S. K., J. L. Hill, D. O. Kleemann, and C. D. Nancarrow. "Development of Ovine Embryos in Synthetic Oviductal Fluid Containing Amino Acids at Oviductal Fluid Concentrations." *Biology of Reproduction* 55 (1996): 703–8.

Wilmut, I., A. E. Schnieke, J. McWhir, A. J. Kind, and K. H. S. Campbell. "Viable Offspring Derived from Fetal and Adult Mammalian Cells." *Nature* 385 (1997): 810–13.

Wilson, F. H., A. Hariri, A. Farhi, et al. "A Cluster of Metabolic Defects Caused by Mutation in a Mitochondrial tRNA." *Science* 306 (2004): 1190–94.

Books

Kolata, Gina. *Clone: The Road to Dolly, and the Path Ahead.* New York: Morrow, 1998.

Nuffield Council on Bioethics. *The Ethics of Research Involving Animals.* London, 25 May 2005.

Wilmut, Ian, Keith Campbell, and Colin Tudge. *The Second Creation: Dolly and the Age of Biological Control.* New York: Farrar, Straus and Giroux, 2000.

Web Sites

Advanced Cell Technology Homepage. http://www.advancedcell.com.

Alexion Pharmaceuticals Inc. Homepage. http://www.alexionpharmaceuticals.com.

Berg, Paul. "Asilomar and Recombinant DNA." Nobel Prize Organization Homepage. http://nobelprize.org.

Biotechnology and Biological Sciences Research Council Homepage. http://www.bbsrc.ac.uk.

Medarex Homepage. http://www.medarex.com.

Pharming Homepage. http://www.pharming.com.

Rifkin, Jeremy. Foundation on Economic Trends Homepage. http://www.foet.org.

Roman Curia Pontifical Academies Homepage. http://www.vatican
.va/roman_curia/pontifical_academies/acdscien.

Royal Society Homepage. http://www.royalsoc.ac.uk.

Society, Religion and Technology Project Homepage. http://www
.srtp.org.uk.

Uncaged Campaigns Xenotransplantation. http://www.uncaged.co.uk/
xeno.htm.

CHAPTER 4: FARMYARD CLONES

Interviews, Correspondences, and Speeches

Christmas, Robin. E-mail correspondence with Roger Highfield, 10
and 12 July 2005.

Colman, Alan. E-mail correspondence with Roger Highfield, 17 and 25
April 2005.

Rosenbaum, Linda. E-mail correspondence with Roger Highfield, 8
July 2005.

Robl, Jim. E-mail correspondence with Roger Highfield, 23 and 24
Feb. and 8 and 13 April 2005.

Tuddenham, Ted. E-mail correspondence with Roger Highfield, 17
March and 4, 5, 8, and 11 July 2005.

Journal and Newspaper Articles

Briggs, R A., A. S. Douglas, R. G. Macfarlane, et al. "Christmas Disease: A Condition Previously Mistaken for Hemophilia." *British
Medical Journal* ii (1952): 1378–82.

Clark, A. J., H. Bessos, J. O. Bishop, et al. "Expression of Human Anti-
hemophilic Factor IX in the Milk of Transgenic Sheep." *Bio/Tech-
nology* 7 (1989): 487–92.

Clark, J., and B. Whitelaw. "A Future for Transgenic Livestock."
Nature Reviews: Genetics 4, no. 10 (2003): 825–33.

Dai, Y., T. D. Vaught, J. Boone, et al. "Targeted Disruption of the

Alpha-1,3-Galactosyltransferase Gene in Cloned Pigs." *Nature Biotechnology* 20 (2002): 251–55.

Fire, A., S. Xu, M. K. Montgomery, S. A. Kostas, S. E. Driver, and C. C. Mello. "Potent and Specific Genetic Interference by Double Stranded RNA in *C. elegans*." *Nature* 391 (1998): 806–11.

Giangrande, Paul. "Six Characters in Search of an Author: The History of the Nomenclature of Coagulation Factors." *British Journal of Hematology* 121 (2003): 703–12.

Hamilton, A. J., and D. C. Baulcombe. "A Novel Species of Small Antisense RNA in Post-transcriptional Gene Silencing." *Science* 286 (1999): 950–52.

Highfield, Roger. "Brave New World for Man and Beast Down on the Farm." *Daily Telegraph*, 6 July 1987, p. 9.

———. "Farm Animals 'Chemical Factories' of the Future." *Daily Telegraph*, 11 Aug. 1987, p. 12.

Lai, L., D. Kolber-Simonds, K. W. Park, et al. "Production of Alpha-1,3-Galactosyltransferase Knockout Pigs by Nuclear Transfer Cloning." *Science* 295 (2002): 1089–92.

Mannucci, P. M., and E. G. Tuddenham. "The Hemophilias: From Royal Genes to Gene Therapy." *New England Journal of Medicine* 344 (2001): 1773–79.

McCreath, K. J., J. Howcroft, K. H. Campbell, A. Colman, A. E. Schnieke, and A. J. Kind. "Production of Gene-Targeted Sheep by Nuclear Transfer from Cultured Somatic Cells." *Nature* 405 (2000): 1066–69.

Schnieke, A. E., A. J. Kind, W. A. Ritchie, et al. "Human Factor IX Transgenic Sheep Produced by Transfer of Nuclei from Trans-fected Fetal Fibroblasts." *Science* 278 (1997): 2130–33.

Wall, R. J., A. M. Powell, M. J. Paape, et al. "Genetically Enhanced Cows Resist Intramammary *Staphylococcus aureus* Infection." *Nature Biotechnology* 23 (2005): 445–51.

Weiss, R. "Xenografts and Retroviruses." *Science* 285 (1999): 1221–22.

Reports

Nuffield Council on Bioethics. 1996. *Animal-to-Human Transplants: The Ethics of Xenotransplantation.* London, 1996.
Institute of Medicine. *Xenotransplantation, Science, Ethics and Public Policy.* Washington, D.C., 1996
Department of Health Advisory Group. *Department of Health Advisory Group on the Ethics of Xenotransplantation Animal Tissues into Humans.* London: Her Majesty's Stationery Office, 1997.

Web Sites

Canadian Hemophilia Society Homepage. http://www.hemophilia.ca.
Geron Homepage. http://www.geron.com.
GTC Biotherapeutics Homepage. http://www.transgenics.com.
Hematech Hompage. http://www.hematech.com.
Nexia Biotechnologies Inc. Homepage. http://www.nexiabiotech.com.

CHAPTER 5: CLONING FOR MY FATHER

Interviews, Correspondences, and Speeches

Andrews, Peter. E-mail correspondence with Roger Highfield, 9 and 10 Feb. 2004 and 25 June 2005.
Blakemore, Colin. Interview with Roger Highfield, UK Stem Cell Bank, South Mimms, 19 May 2004 and 18 Jan. 2005.
Cui, Wei. Roslin Institute press briefing, 3 March 2005.
De Sousa, Paul. British Association for the Advancement of Science, Dublin, 9 Sept. 2005; Roslin Institute press briefing, 3 March 2005.
———. E-mail correspondence with Roger Highfield, 4 June 2004 and 5 and 7 March 2005.
Hwang, W. S. E-mail correspondence with Roger Highfield, 10 Feb. 2004 and 21 May, 29 July 29, and 2 Aug. 2005.
———. Press conferences at the American Association for the Advancement of Science, Seattle, 12 Feb. 2004.

————. Press conferences with Gerald Schatten at Royal Institution, London, 19 May 2005.

Lovell-Badge, Robin. Interview with Roger Highfield, Science Media Center, London, 31 May 2005.

McWhir, Jim. Roslin Institute press briefing, 3 March 2005.

Minger, Stephen. Interview with Roger Highfield, London, 12 Jan. 2005.

Murdoch, Alison. E-mail correspondence with Roger Highfield, 13 and 25 May, 15 and 24 June, 18 and 27 Aug. 2004 and 1 Feb. and 19 May 2005.

————. Interview with Roger Highfield, Centre for Life, Newcastle-upon-Tyne, 19 Oct. 2004.

Pedersen, Roger. Interview with Roger Highfield, Science Media Centre, London, 31 May 2005.

Schatten, Gerald. E-mail correspondence with Roger Highfield, 8 Jan. 2006.

Shaw, Christopher. Interview with Roger Highfield, Science Media Centre, London, 28 Sept. 2004.

Smith, Austin. E-mail correspondence with Roger Highfield, 23 Nov. 2004 and 19 Jan., 3 March, 27 June, and 15, 16, 22, 23, and 25 Aug. 2005.

————. Interview with Roger Highfield, The Tun, Edinburgh, 3 March 2005.

Stojkovic, Miodrag. E-mail correspondence with Roger Highfield, 15 and 24 June, 28 July, 22 Sept., 13 Oct., 22 Nov., and 22 Dec. 2004 and 13, 18, and 19 April and 19 May 2005.

————. Interview with Roger Highfield, Centre for Life, Newcastle-upon-Tyne, 19 Oct 2004, and by phone, 10 Jan. 2006.

Journal and Newspaper Articles

Faiola, Anthony, and Rick Weiss. "South Korean Panel Debunks Scientist's Stem Cell Claims: Fraud Finding Is Another Setback in Cloned Embryo Research," *Washington Post*, 10 Jan. 2006, p. A09.

Highfield, Roger. "Finding the Wonder Cell." *Daily Telegraph*, 25 July 1988, p. 8.

———. "Have We Been Oversold the Stem Cell Dream?" *Daily Telegraph*, 29 June 2005, p. 21.

———. "Scientists Take a Giant Step Forward in Human Cloning." *Daily Telegraph*, 20 May 2005, p. 1.

Highfield, Roger. "New Doubts Cast on Human Clone Claim." *Daily Telegraph*, 15 Dec. 2005, p. 10.

———. "Cloning Star Who Fooled the World." *Daily Telegraph*, 24 Dec. 2005, p. 4.

———. "Embryo Cloning Cheat Resigns in Disgrace." *Daily Telegraph*, 24 Dec. 2005, p. 1.

———. "Disgraced Cloning Expert Blames Colleague." *Daily Telegraph*, 27 Dec. 2005, p. 2.

———. "Clone Doctor Faked His Work on Stem Cells." *Daily Telegraph*, 11 Jan. 2006, p. 14.

Highfield, Roger, and David Derbyshire. "Human Cells Cloned: Babies Next?" *Daily Telegraph*, 13 Feb. 2004, p. 1.

Hwang, W. S., Y. J. Ryu, J. H. Park, et al. "Evidence of a Pluripotent Human Embryonic Stem Cell Line Derived from a Cloned Blastocyst." *Science* 303 (2004): 1669–74.

Hwang, W. S., S. I. Roh, B. C. Lee, et al. "Patient-Specific Embryonic Stem Cells Derived from Human SCNT Blastocysts." *Science* 308 (2005): 1777–83.

Jiang, Y., B. N. Jahagirdar, R. L. Reinhardt, et al. "Pluripotent Nature of Adult Marrow Derived Mesenchymal Stem Cells." *Nature* 418 (2002): 41–49.

Kono, T., Y. Obata, Q. Wu, et al. "Birth of Parthenogenetic Mice That Can Develop to Adulthood." *Nature* 428 (2004): 860–64.

Marshall, E., and G. Vogel. "Cloning Announcement Sparks Debate and Scientific Skepticism." *Science* 294 (2001): 1802–3.

Martin, M., A. Muotri, F. Gage, and A. Varki. "Human Embryonic Stem Cells Express an Immunogenic Nonhuman Sialic Acid." *Nature Medicine* 11 (2005): 228–32.

Park, Alice, and Stella Kim. "Independent Panel Confirms Stem Cell Fraud. Outside Investigators Determine that 2004 Cloned Stem Cell Lines from South Korean Lab Were Faked." *Time Online*, 9 Jan. 2006.

Pickering, S. J., S. L. Minger, M. Patel, et al. "Generation of a Human Embryonic Stem Cell Line Encoding the Cystic Fibrosis Mutation ΔF508, Using Preimplantation Genetic Diagnosis." *Reproductive BioMedicine Online* 10 (2005): 390–97.

Samuel Reich, Eugenie. Stem-cell Pioneer's Findings in Doubt: Woo Suk Hwang Is in Trouble Again, This Time over His Claims to Have Made Personalised Cloned Cells." *New Scientist* 188, no. 2530 (17 Dec. 2005), p. 6.

Shamblott, M. J., J. Axelman, S. Wang, et al. "Derivation of Pluripotent Stem Cells from Cultured Human Primordial Germ Cells." *Proceedings of the National Academy of Sciences* 95 (1998): 13726–31.

Simerly, C., T. Dominko, C. Navara, et al. "Molecular Correlates of Primate Nuclear Transfer Failures." *Science* 300 (2003): 297.

Skloot, R. "Henrietta's Dance." *Johns Hopkins Magazine Online*, April 2000.

Stojkovic, M., P. Stojkovic, C. Leary, V. J. Hall, L. Armstrong, M. Herbert, M. Nesbitt, M. Lako, A. Murdoch. "Derivation of a Human Blastocyst after Heterologous Nuclear Transfer to Donated Oocytes." *RBM Online* 11 (2005): 2: 226–31.

Thomson, J. A., J. Itskovitz-Eldor, S. S. Shapiro, et al. "Embryonic Stem Cell Lines Derived from Human Blastocysts." *Science* 282 (1998): 1145–47.

Ying, Q. L., J. Nichols, E. P. Evans, and A. G. Smith. "Changing Potency by Spontaneous Fusion." *Nature* 416 (2002): 545–48.

Reports

Chief Medical Officer's Expert Group on Therapeutic Cloning. *Stem Cell Research: Medical Progress with Responsibility*. United Kingdom Department of Health, 2000.

House of Lords Select Committee. *Stem Cell Research*. London, 27 Feb. 2002.

Human Fertilisation and Embryology Authority/Human Genetics Advisory Committee. *Cloning Issues in Reproduction, Science and Medicine*. London, 1998.

Royal Society. *Whither Cloning?* London, 1 Jan. 1998.

Seoul National University Investigation Committee. *Summary of the Final Report on Professor Woo Suk Hwang's Research Allegations.* 9 Jan. 2006.

Warnock, Dame Mary. *Report of the Committee of Inquiry into Human Fertilisation and Embryology.* London: Her Majesty's Stationery Office, 1984.

Web Sites

Advanced Cell Technology Homepage. http://www.advancedcell.com.

Center for Life Homepage. http://www.nfc-life.org.uk.

Comment on Reproductive Ethics Homepage. http://www.corethics .org.

Human Fertilisation and Embryology Authority Homepage. http:// www.hfea.gov.uk.

Human Genetics Commission Homepage. http://www.hgc.gov.uk.

Nuffield Council on Bioethics Homepage. http://www.nuffield.org .uk/bioethics.

Royal Society Homepage. http://www.royalsoc.ac.uk/policy.

U.K. Stem Cell Bank Homepage. http://www.ukstemcellbank.co.uk.

Stem cell research centers/information sites include these: http://www.iscr.ed.ac.uk, http://www.stemcells.cam.ac.uk, http:// www.snu.ac.kr, http://www.stemcell.harvard.edu, http://stem cells.nih.gov, http://www.nscc.edu.au, http://www.stemcell forum.org, and http://www.isscr.org.

Unpublished Material

Hwang, W. S. Notes handed to journalists at the American Association for the Advancement of Science, 12 Feb. 2004.

Pedersen, R. "Stem Cells 2005: Progress to Therapy." Edinburgh International Conference Center, 4–5 March 2005.

CHAPTER 6: IS A BLASTOCYST A PERSON?

Interviews, Correspondences, and Speeches

Banner, Michael. E-mail correspondence with Roger Highfield, 24 March and 7 April 2005.

———. Interview with Roger Highfield, 18 March 2005.

Cibelli, Jose. E-mail correspondence with Roger Highfield, 7 Jan. 2005.

Edwards, Robert. Press conference at Sixth PGD Symposium, Queen Elizabeth II Conference Center, London, 19 May 2005.

Harris, John. E-mail correspondence with Roger Highfield, 15 and 29 March and 16 and 17 May 2005.

———. Interview with Roger Highfield, Manchester, 24 and 25 Feb. 2005.

Life. Press Release. 27 Feb. 2002.

Zernicka-Goetz, Magda. E-mail correspondence with Roger Highfield, 20 May and 10 and 22 June 2005.

Journal and Newspaper Articles

Allen, Mike. "Destruction of Life Is Part of Culture of Death." *Lexington Herald Leader*, 6 Sept. 2005, p. A6.

Cleland, Carol E., and Christopher F. Chyba. "Defining Life." *Origins of Life and Evolution of the Biosphere* 32 (2002): 387–93.

Connor, Kenneth. "Guard Sanctity of Life." *USA Today*, 10 Aug. 2001, p. 10A.

Cookson, Clive, and Daniel Green. "A Wolf in Sheeps' Clothing?" *Financial Times*, 6 Dec. 1997, p. 6.

Editorial. "Bush Advocates Smartest Position on Stem Cells." *Buffalo News*, 15 Aug. 2004, p. H5.

Editorial. "Is It Wrong to Clone Humans?" *Express*, 13 Feb. 2004, p. 12.

Greene, M., K. Schill, S. Takahashi, et al. "Moral Issues of Human–Non-Human Primate Neural Grafting." *Science* 309 (2005): 385–86.

Harper, J. C., and J. D. A. Delhanty. "Preimplantation Genetic Diagnosis: Current Opinions in Obstetrics and Gynaecology." *Highfielder* 12 (April 2000): 67–72.

Steptoe, P., and R. Edwards. "Birth after the Reimplantation of a Human Embryo." *Lancet* 2, no. 8085 (1978): 366.

Timmons, Heather. "U.K. Clears Cloning of Human Embryos." *International Herald Tribune*, 12 Aug. 2004, p. 1.

Books

Barrow, John. *Pi in the Sky: Counting, Thinking, and Being.* Oxford: Clarendon, 1992.

Coveney, Peter, and Roger Highfield. *Frontiers of Complexity: The Search for Order in a Chaotic World.* London: Faber, 1995.

Harris, John. *The Value of Life.* London: Routledge, 1985.

Klotzko, Arlene Judith. *A Clone of Your Own?: The Science and Ethics of Cloning.* Oxford: Oxford University Press, 2004.

Schrödinger, Erwin. *What Is Life? With Mind and Matter and Autobiographical Sketches.* Cambridge: Canto, 1992.

Reports

Human Genetics Advisory Commission and Human Fertilisation and Embryology Authority. *Cloning Issues in Reproduction, Science and Medicine.* London, Jan. 1998.

Medical Research Council (MRC). *Report of the MRC Expert Group on Fetal Pain.* London, 28 Aug. 2001.

Web Sites

Comment on Reproductive Ethics. http://www.corethics.org.

Eurostem. http://www.eirma.asso.fr/eurostem.html.

Life, the UK's Leading Pro-life Charity. http://www.lifeuk.org.

Motor Neurone Disease Association. http://www.mndassociation.org.

National Right to Life. http://www.nrlc.org.

Nuffield Council on Bioethics. http://www.nuffieldbioethics.org.

President George W. Bush. http://www.whitehouse.gov.

Select Committee on Science and Technology. http://www.parliament.uk.

CHAPTER 7: WHY WE SHOULD
NOT CLONE BABIES

Interviews, Correspondences, and Speeches

Collas, Philippe. E-mail correspondence with Roger Highfield, 5 and 6 Jan. 2005.

Ding, Sheng. E-mail correspondence with Roger Highfield, 7 Jan. 2005.

Krawetz, Stephen. E-mail correspondence with Roger Highfield, 5 and 7 April and 2 Aug. 2005.

Lai, Tony. E-mail correspondence with Roger Highfield, 13 Nov. 2002.

Lanza, Robert. E-mail correspondence with Roger Highfield. 23 Sept. 2004 and 5 Jan. and 29 June 2005.

Renard, Jean-Paul. E-mail correspondence with Roger Highfield, 6 April 2005.

Skok, Jane. E-mail correspondence with Roger Highfield, 10, 11, 14, and 15 March 2005.

Young, Lorraine. E-mail correspondence with Roger Highfield, 18 Oct. 2004.

Journal and Newspaper Articles

Dawkins, Richard. "The Secret Confessions of a Closet Clone." *Evening Standard*, 25 Feb. 1997, p. 9.

De Robertis, E. M., G. A. Partington, R. F. Longthorne, and J. B. Gurdon. "Somatic Nuclei in Amphibian Oocytes: Evidence for Selective Gene Expression." *Journal of Embryology and Experimental Morphology* 40 (1977): 199–214.

Heyman Y., P. Chavatte-Palmer, D. LeBourhis, S. Camous, X. Vignon, and J. P. Renard. "Frequency and Occurrence of Late-Gestation

Losses from Cattle Cloned Embryos." *Biology of Reproduction* 66 (2002): 6–13.

Highfield, Roger. "Check on Gene Therapy after Leukemia Scare in France." *Daily Telegraph*, 4 March 2005, p. 11.

———. "The Holy Grail of Stem-Cell Research. Dismantling Human Embryos for Cell Therapy Has Prompted a Clash of Strong Views. But What Are the Alternatives?" *Daily Telegraph*, 19 Jan. 2005, p. 16.

———. "Human Eggs Can Be Mass Produced from Male Embryos." *Daily Telegraph*, 2 May 2003, p. 11.

———. "Scientist Predicts End of Infertility within 10 Years." *Daily Telegraph*, 25 July 2003, p. 6.

Krawetz, S. "Paternal Contribution: New Insights and Future Challenges." *Nature Reviews: Genetics* 6 (2005): 633–42.

Lanza, R., J. B. Cibelli, C. Blackwell, et al. "Extension of Cell Life-span and Telomere Length in Animals Cloned from Senescent Somatic Cells." *Science* 288 (2000): 665–69.

Peura, T. "Improved *In Vitro* Development Rates of Sheep Somatic Nuclear Transfer Embryos by Using a Reverse-Order Zona-Free Cloning Method." *Cloning and Stem Cells* 5 (2003): 13–24.

Ribas, R., B. Oback, W. Ritchie, et al. "Development of a Zona-Free Method of Nuclear Transfer in the Mouse." *Cloning and Stem Cells* 7 (2005): 126–38.

Rideout, W. M., K. Eggan, and R. Jaenisch. "Nuclear Cloning and Epigenetic Reprogramming of the Genome." *Science* 293 (2001): 1093–98.

Sakatani, T., A. Kaneda, C. A. Iacobuzio-Donahue, et al. "Loss of Imprinting of *Igf2* Alters Intestinal Maturation and Tumorigenesis in Mice." *Science* 307 (2005): 1976–78.

Wakayama, T., Y. Shinkai, K. L. Tamashiro, et al. "Cloning of Mice to Six Generations." *Nature* 407 (2000): 318–19.

Wells, D. N., B. Oback, and G. Laible. "Cloning Livestock: A Return to Embryonic Cells." *Trends in Biotechnology* 21 (2003): 428–32.

Wells, D. N., J. T. Forsyth, V. McMillan, and B. Oback. "The Health of Somatic Cell Cloned Cattle and Their Offspring." *Cloning Stem Cells* 6 (2004): 101–10.

Zhou, Q., J. P. Renard, G. Le Friec, et al. "Generation of Fertile Cloned Rats by Regulating Oocyte Activation." *Science* 302 (2003): 1179.

Books

Levick, Stephen E. *Clone Being: Exploring the Psychological and Social Dimensions*. Lanham, Md.: Rowman and Littlefield, 2004.

Web Sites

AgResearch Homepage. http://www.agresearch.co.nz.
British Fertility Society Homepage. http://www.britishfertilitysociety .org.uk.
National Institute for Agricultural Research (INRA) Homepage. http://www.inra.fr/english.

Unpublished Material

Brown, Louise. "The Baby Makers." London, Channel 4, 1999.

CHAPTERS 8 AND 9: DESIGNER BABIES AND BEYOND HUMAN CLONING

Interviews, Correspondences, and Speeches

Braude, Peter. Interview with Roger Highfield, Science Media Center, London, 25 April 2005.
Delhanty, Joy. E-mail correspondence with Roger Highfield, 6, 14, 22 and 25 Jan. and 31 March 2005.
Fischer, Alain. E-mail correspondence with Roger Highfield, 26 April 2000, 9 Oct. 2004, and 2 March 2005.
Gaspar, Bobby. Interview with Roger Highfield, London, 26 June 2003.
Handyside, Alan. Press conference, London, 18 April 1990.
Thrasher, Adrian. E-mail correspondence with Roger Highfield, 1 Nov.

and 7 Oct. 2004 and 25 Feb., 1, 3, 9, 11, and 30 March, 22 April, and 30 June 2005.

———. Interview with Roger Highfield, London, 26 June 2003.

Turnbull, Doug. E-mail correspondence with Roger Highfield, 15 March, 15 and 30 Aug., and 2, 4, 7, and 8 Sept. 2005.

Watson, James. Press conference, Royal Society, London, 23 April 2003.

———. Press conference, Sixth PGD Symposium, Queen Elizabeth II Conference Center, London, 19 May 2005.

Winston, Robert. Interviews with Roger Highfield, London, 18 April 1990 and 26 Jan. 2000; Dublin, 9 Sept. 2005.

Journal and Newspaper Articles

Barritt, J., S. Willadsen, C. Brenner, and J. Cohen. "Cytoplasmic Transfer in Assisted Reproduction." *Human Reproduction 7* (2001): 428–35.

Chan, A. W. S., K. Y. Chong, C. Martinovich, C. Simerly, and G. Schatten. "Transgenic Monkeys Produced by Retroviral Gene Transfer into Mature Oocytes." *Science* 291 (2001): 309–12.

Boyle, R., and J. Savulescu. "Ethics of Using Preimplantation Genetic Diagnosis to Select a Stem Cell Donor for an Existing Person." *British Medical Journal* 323 (2001): 1240–43.

Hacein-Bey-Abina, S., C. Von Kalle, M. Schmidt, et al. "*LMO2*-Associated Clonal T Cell Proliferation in Two Patients after Gene Therapy for SCID-X1." *Science* 302 (2003): 415–19.

Handyside, A. H., E. H. Kontogianni, K. Hardy, and R. M. L. Winston. "Pregnancies from Biopsied Human Preimplantation Embryos Sexed by Y-Specific DNA Amplification." *Nature* 344 (1990): 768–70.

Harris, J. "Intimations of Immortality." *Science* 288 (2000): 59.

Highfield, Roger. "The Beginning of the End for Fatal Hereditary Disease." *Daily Telegraph*, 18 Jan. 1988, p. 12.

———. "British Team Discovers Gene Lifeline for Children." *Daily Telegraph*, 28 June 2003, p. 15.

———. "The Ethics of Having a Baby as a Spare Part." *Daily Telegraph*, April 9 1990, p. 16.

———. "How Do You Define a Human Being?" *Daily Telegraph*, 28 Oct. 1991, p. 12.

———. "Is There a Case for Modifying Her Genes?" *Daily Telegraph*, 31 Aug. 2005, p. 14.

———. "Meet the Baby with Three Parents." *Daily Telegraph*, 30 May 2001, p. 25.

———. "Researchers Open the Way to Screening Test-Tube Babies for Genetic Diseases." *Daily Telegraph*, 19 April 1990, p. 1.

———. "Scientists Call for Limits to 'Designer Baby' Transplants." *Daily Telegraph*, 3 June 1988, p. 7.

———. "Why Life Will Never Be the Same Again. The New Genetics." *Telegraph Magazine*, 17 Aug. 1991, p. 26.

Slater, Eve. "Today's FDA." *New England Journal of Medicine* 352 (2005): 293.

Thrasher, A. J. "Gene Therapy: Great Expectations?" *Medical Journal of Australia* 182 (2005): 440–41.

Wells, D., and J. Delhanty. "Comprehensive Chromosomal Analysis of Human Preimplantation Embryos Using Whole Genome Amplification and Single Cell Comparative Genomic Hybridization." *Molecular Human Reproduction* 6 (2000): 1055–62.

Williams, T., T. W. Mwangi, D. J. Roberts, et al. "An Immune Basis for Malaria Protection by the Sickle Cell Trait." *Public Library of Science Medicine* 2, no. 5 (2005): 128.

Books

Andrews, Lori. *The Clone Age*. New York: Owl Books, 2000.

Frankel, Mark, and Audrey Chapman. *Human Inheritable Genetic Modifications*. Washington, D.C.: AAAS, 2000.

Klotzko, Arlene Judith. *A Clone of Your Own?: The Science and Ethics of Cloning*. Oxford: Oxford University Press, 2000.

Levick, Stephen E. *Clone Being: Exploring the Psychological and Social Dimensions*. Lanham, Md.: Rowman and Littlefield, 2004.

Silver, Lee M. *Remaking Eden: Cloning and Beyond in a Brave New World*. New York: William Morrow, 1997.

Watson, James. DNA: *The Secret of Life*. New York: Alfred A. Knopf, 2003.

Web Sites

American Society for Reproductive Medicine Homepage. http://www.asrm.org.

American Society of Gene Therapy Homepage. http://www.asgt.org.

European Society of Human Reproduction and Embryology Homepage. http://www.eshre.com.

Huntington's Disease Association Homepage. http://www.hda.org.uk.

Sickle Cell Society Homepage. http://www.sicklecellsociety.org.

GLOSSARY

Terms shown in **bold** are explained elsewhere in the glossary.

activation   The process during **fertilization** by which the sperm induces the egg (**oocyte**) to resume its development. This triggers the completion of the second phase of **meiosis**, the process by which the egg halves its number of **chromosomes**. The activation caused by the sperm can be mimicked by treating the egg with selected chemicals or applying an electric shock.

amino acids   The molecular building blocks of **proteins**.

asexual reproduction   Reproduction that does not require the union of egg and sperm, where all the genetic material of an offspring comes from a single progenitor. In other words, cloning.

atoms   The building blocks of all the molecules in living things. Atoms were thought by the ancient Greeks to be indivisible units of matter. Now they are seen as the smallest units that bear the chemical characteristics of an element, whether hydrogen or uranium. Around 200,000 million million atoms would fit on the period at the end of this sentence. They are mostly empty space: the **nucleus**, where most mass resides, is 100,000 times smaller than the overall atom. Each atom consists of a positively charged nucleus orbited by a mist of negative charge.

base
The molecular units—"letters" of genetic code, called nucleotides—that provide the variation in a strand of DNA or RNA. DNA has four different types of base—thymine, cytosine, adenine, and guanine—while RNA contains uracil instead of thymine. The two molecular "letters" on opposite—complementary—DNA or RNA strands are called a base pair.

blastocyst
A fertilized egg when it reaches the stage of a microscopic ball of approximately 40 to 200 cells, sometimes referred to as a pre-embryo. The cells form a hollow sphere, comprising an **inner cell mass** consisting of embryonic stem cells that will become the fetus surrounded by an outer ring of cells—the trophectoderm—that will become part of the **placenta**.

cell
The building block of bodies. A discrete, membrane-bound portion of living matter, the smallest unit capable of an independent existence. Complex organisms such as humans are made up of more than two hundred different types, such as muscle, bone, and nerve. Overall, there are two basic types—somatic (body) cells and **germ line** (reproductive) cells: eggs and sperm. Forty human cells placed in a line would add up to around a millimeter.

cell cycle   The life story of a cell, which can be divided into a period of growth during which the chromosomal **DNA** is copied and cell division (**mitosis** or **meiosis**) occurs. The story consists of various chapters: the phase when DNA is replicated (S phase), the phase when the cell actually divides into two cells (M phase), the two intervening gap phases (G1 and G2), and a nondividing state called **quiescence** (G0).

cell division   During normal human cell division, the **chromosomes** align neatly in the middle of the cell into twenty-three pencil-shaped pairs of parallel strands called **chromatids**. At the same time two **centrioles** form at opposite poles of the cell. Threadlike **microtubules** attach each of the chromatids to the centrioles, forming what is called the **spindle**. When the cell divides, the chromatids separate and are pulled by the microtubules to opposite poles. Thus, two identical pools of chromatids, the inherited genetic instructions of the future daughter cells, form at the poles. The interior of the cell separates in barbell-like fashion with a narrow bridge that eventually pinches in two to form the two daughter cells.

centrioles   Cellular organelles adjacent to the **nucleus** that help form the **spindle** during cell division, and consist of a cylinder with nine microtubules arranged peripherally in a circle.

chimera   **Embryos** or offspring made up of cells from more than one embryo. The resulting animals are blends of cells from what would normally be two separate animals, each with its unique **DNA**.

chromatid   One of two identical strands into which a **chromosome** splits during **mitosis**, or cell division.

chromatin   That portion of the cell nucleus that contains all of the
              DNA of the **nucleus** in animal or plant cells. When
              cells divide, the chromatin is seen as distinct **chromo-
              somes**, duplicating, with an equal partition of each set
              of chromosomes then traveling to each of the new
              daughter cells.

chromosome   Carrier of genetic information in the form of **DNA**
              that has been tightly wrapped up, with specialized
              proteins such as **histones**, to form a package. This
              package, the chromosome, is the visible form that
              DNA takes within the cell. Usually an animal's DNA
              is stored in many chromosomes, like volumes of an
              encyclopedia. Human cells have forty-six chromo-
              somes. There are twenty-three pairs in all human cells,
              and half that number in eggs and sperm.

cleavage   The process of cell division that allows the **zygote** (the
              newly fertilized egg) to develop into a **blastocyst**.

clone   Clones are genetically identical individuals produced
              by a form of **asexual reproduction**. Plant cuttings are
              clones. In the context of this book "clone" is used to
              describe offspring produced by transfer of the nucleus
              from one cell into an egg (**oocyte**) from which the
              genetic information has been removed: the process of
              nuclear transfer.

cytoplasm   Cells are composed of an inner sanctum called the
              nucleus, which contains the **DNA** and which resides
              within a surrounding cytoplasm, bound by the cell
              wall, within which many processes occur. The cyto-
              plasm is highly organized and contains specialized
              structures called **organelles**.

cytoskeleton The network of fibers that form the "skeleton" of a cell.

culture
: The term used by scientists to describe the feeding of cells and keeping them alive in the laboratory. Typically, cells and embryos are grown in a special liquid, known as a culture medium, which provides all of the energy and other resources that are required for the well-being of the cells.

diabetes
: A disorder of the metabolism that strikes when the body cannot balance the storage and use of energy because of an error in either the production of the hormone insulin or the response to insulin. There are two main types of diabetes. In type I, or insulin-dependent, diabetes the patient does not produce adequate amounts of insulin, because insulin-producing cells in the pancreas—"islet cells"—have been destroyed by the body's immune system. Such patients respond well to injections of insulin. By contrast, in type II diabetes the person no longer makes enough insulin or no longer responds to injected insulin. Type II is the more common form, usually occurs in adults, and is linked to obesity.

differentiation
: The process by which a primitive **stem cell** from an embryo "commits" to becoming a specialized cell in the body, such as a skin cell or a bone cell.

diploid
: Cells differ in the number of copies of each **chromosome** they contain. Diploid cells contain two sets of chromosomes, whereas **gametes**—eggs and sperm—contain only one set and are described as **haploid**. A diploid human cell contains forty-six chromosomes; a haploid egg or sperm, twenty-three.

DNA
: Short for **deoxyribonucleic** acid, the vehicle of inheritance for all creatures. The complex giant nucleic acid molecule carries the genetic recipe for

the design and assembly of proteins, the basic building blocks of life. The chainlike DNA molecule is in the form of a double helix made up of a series of "bases," which come in four types (adenine, guanine, cytosine, and thymine, or A, G, C, and T). The order of these bases provides a recipe (first issued through the medium of another molecule, **RNA**) for the proteins that are the building blocks of life. The code is a three-letter one, with a triplet of letters (ATT, say) coding for a particular **amino acid** that, when joined with a string of others, makes a **protein**. Each person has about three billion bases in his genetic makeup, or **genome**, but only around 25,000 genes that work to make proteins.

dopamine A substance found in the brain that transmits messages between nerves. **Parkinson's disease**, the brain disorder, is a result of inadequate functioning of dopamine-producing **neurons** in a part of the brain called the substantia nigra. Dopamine allows smooth, coordinated function of the body's muscles and movement. When approximately 80 percent of the dopamine-producing cells are damaged, the symptoms of Parkinson's disease appear, such as tremor and slowness of movement.

ectoderm The part of a developing **embryo** that gives rise to epidermis—the outer layer of the skin—and nerve cells. It is one of three "germ layers," the basic three cellular layers—ectoderm, **endoderm**, and **mesoderm**—from which the organs and tissues of the body develop through further differentiation.

embryo An organism in the early stages of growth and differentiation. During sexual reproduction it formed at **fertilization** by the union of sperm and egg.

embryonic stem cell   The early **embryo** contains **stem cells**. Stem cells have the characteristic of being able to divide many times and give rise to the two hundred or so cell types in a body. Embryonic stem cells retain the ability to form all of the different tissues of an adult.

endoderm   That part of an **embryo** that gives rise to the epithelium (lining) of the gut and gut derivatives. It is one of three germ layers. See also **ectoderm**.

enucleate   To remove a nucleus from a cell, usually an egg (**oocyte**), in a process carried out during **nuclear transfer**. However, the term is somewhat misleading: the egg cells often used in cloning are in a special stage called MII oocytes and don't actually have a **nucleus**, but rather have free **chromosomes**.

enzyme   The word "enzyme," which denotes the huge **proteins** that cells use to transform other molecules, was coined from the Greek words for "in yeast." It refers to a biological catalyst usually composed of a large protein molecule consisting of many, even thousands, of **atoms** that accelerates essential chemical reactions in living cells. While a cell has a size of the order of micrometers (millionths of a meter), enzymes and other proteins have a size of several nanometers (billionths of a meter). Ten hydrogen atoms laid side by side can cover a nanometer, which is one-thousandth the length of a typical bacterium and one-millionth the size of a pinhead.

epigenetic modification   The process of turning **genes** on and off during cell differentiation. One way to achieve it is to alter the way **DNA** is decorated with methyl groups, a process called DNA methylation. Another is through the action of **histones**.

epigenetic The process of removing **epigenetic modifications**
reprogramming of DNA, so that genes whose expression was turned
off during embryonic development and cell differentiation can become active again. In other words, turning
back the developmental clock so that the epigenetic
modifications that turned an embryonic stem cell into
a liver cell, for example, are removed. With the developmental slate wiped clean, this cell can once again be
persuaded to develop into various types.

epigenetics Changes in how **genes** are used in the body that take
place without changes to the genes themselves,
notably during development. It is these epigenetic
changes that make, say, a lung cell so distinctive from
a brain cell, even though they contain the same **DNA**.
One way to achieve epigenetic modification is to alter
the way DNA is decorated with methyl groups, a
process called DNA methylation. Another is through
the action of **histones**. In **epigenetic reprogramming**,
genes that were turned off during embryonic development and cell differentiation can become active again.
In other words, the developmental clock is wound
back so that the epigenetic modifications that turned
an **embryonic stem cell** into a liver cell, for example,
are removed. With the developmental slate wiped
clean, this cell can once again be persuaded to develop
into various types.

eugenics An attempt to alter (with the aim of "improving") the
genetic constitution of future generations by control-
ling reproduction. Popular in the first part of the
twentieth century, when individuals with traits considered desirable were encouraged to reproduce; those
with traits considered undesirable were discouraged
from having children. Thousands were involuntarily
sterilized. Eugenic ideologies in the United States, for

example, resulted in social legislation to keep racial and ethnic groups separate, to restrict immigration from southern and eastern Europe, and to sterilize people considered genetically unfit. The ideas of the eugenics movement were models for the Nazis, whose eugenic ideologies culminated in the Holocaust.

evolution From the Latin *evolutio*, "unfolding." The idea of shared descent of all creatures, men, sheep, and mice included. The names of Charles Robert Darwin (1809–1882) and Alfred Russel Wallace (1823–1913) have long been joined with the modern concept of evolution and the theory of natural selection. In his *The Origin of Species* (1859), Darwin suggested a mechanism: inherited diversity, a struggle for existence that means that not all those born can survive and pass on their heritage; and natural selection (inherited differences in the chance of reproduction). Variants that increase their carrier's ability to make copies of themselves hence become more common; those that hinder it, more rare. In time, Darwin suggested, this led to the evolution of new forms of living organism—the origin of species. His ideas fit perfectly with those of modern genetics. Diversity arises through mutation, random changes in the genetic material **DNA**, from generation to generation.

fertilization The fusion of sperm with egg, during which the sperm contributes genetic material to the egg (**oocyte**) and stimulates activation of that oocyte.

fibroblast A cell commonly grown in the laboratory because of the ease with which it can be cultured and made to proliferate.

gamete A reproductive cell (egg or sperm).

gene The unit of heredity consisting of a **DNA** sequence that directs production of a particular protein. Everyone inherits two copies of each gene, one from each parent. A dominantly inherited genetic disease occurs when only one copy of the gene is sufficient to produce the disease, for example Huntington's chorea. A recessively inherited disease occurs only if both copies of the defective gene are present, as happens with Tay-Sachs disease, cystic fibrosis, and sickle-cell disease.

gene expression Genes that are actually in use in the body are said to be "expressed."

gene therapy The transfer of **DNA**—either packed in **viruses**, in fat particles, or even as naked DNA—to treat a genetic disorder. Clinical trials of gene therapy have been under way since 1990. Though results have been disappointing so far, there are several cases now where it has been shown to work. The hope is that it will eventually facilitate treatment of many diseases.

genetic code The sequence of chemical building blocks of **DNA** (bases) which spells out the instructions to make **amino acids,** the molecular building blocks of **proteins.**

genetic engineering Tinkering with the **genetic code**—in the form of DNA—of a creature to produce animals and plants with desirable properties.

genome The entire **DNA** sequence of an organism, consisting of a series of "letters" of **genetic code** (A, C, G, and T). The human genome is around three billion letters.

genomics The science of identifying the sequence of **DNA**—that is, the order of "letters" (chemicals called bases, A, C,

G, and T) in various species, and subsequent processing of that information.

germ cells   Cells that give rise to egg or sperm.

germ line   The lineage of cells that connect our generation to the previous one and even to the first generation of life four billon years ago. The sperm and the egg are examples of germ line cells.

haploid human cell   A cell such as an egg or sperm that contains only twenty-three chromosomes, half the number of the other (**diploid**) cells in the body.

HFEA   Human Fertilisation and Embryology Authority, which at present regulates in the UK all procedures with human **embryos**, whether for research or therapy.

histone   Histones, which are **proteins**, bind to **DNA** and wrap the genetic material into "beads on a string," in which DNA (the string) is wrapped around small blobs of histones (the beads) at regular intervals. Histones play a role in **chromosome** architecture and influence the way that genes are used, or expressed.

implantation   The process whereby an **embryo**, after traveling through the fallopian tube to the **uterus**, embeds itself in the lining of the uterus so that development can continue.

imprinted gene   A family of genes that are used only if they are inherited from the mother or from the father. We have two copies of most genes; exceptions include the genes that determine gender in males. Usually the two copies are used (expressed) equally in the body. In the case of imprinted genes only the copy inherited from one parent is expressed. **Epigenetic** mechanisms control imprinting.

inner cell mass
A clump of cells growing within and to one side of the **blastocyst** from which the **embryo** develops.

insulin
A **protein** hormone that is released into the circulation by the pancreas, which regulates the concentration of glucose in the blood. Diabetes is caused either by a lack of insulin or by lack of a response to insulin.

in vitro
Performed outside the body, that is, in the laboratory (the literal meaning of "*in vitro*" is "in glass"; for example, in a test tube).

in vitro fertilization (IVF)
The union of an egg and sperm, "in glass," where the event takes place outside the body and in an artificial environment such as a test tube or, more commonly, in a Petri dish.

in vivo
Performed in the body.

karyoplast
In effect, the genetic makeup that is used in the **nuclear transfer** cloning process. The karyoplast consists of a donor **nucleus**, with a greater or lesser amount of **cytoplasm** attached, which is transferred to an enucleated egg (**oocyte**) in nuclear transfer.

karyotype
The microscopic appearance of a set of **chromosomes**, including their number, shape, and size.

large-offspring syndrome
A condition often seen in cloned cattle and sheep in which fetuses grow much larger than their normal size in the **uterus** and often suffer organ defects. The syndrome poses serious health risks to the mother and to

the offspring, often resulting in the death of one or
both.

meiosis   The formation of **germ cells**, or **gametes**—eggs and
                 sperm—where the **chromosome** number is cut by
                 half. At the end of this specialized form of **cell divi-
                 sion**, which can take decades in the case of eggs
                 (**oocytes**), the gametes have only half the number of
                 chromosomes of the source cells.

mesoderm   The part of the **embryos** that gives rise to the muscu-
                 lar and circulatory systems and most of the skeletal
                 and urogenital systems. Mesoderm is one of three
                 germ layers. See also **ectoderm** and **endoderm**.

metaphase   The phase of **cell division** when all of the **chromosomes**
                 are assembled by the **cytoskeleton** prior to cell division.

microtubules   A component of the **cytoskeleton**. They have a partic-
                 ular role in separating the pairs of **chromosomes**
                 during **cell division**. In the egg at metaphase II of
                 meiosis, the chromosomes are held on a **spindle** ready
                 for division.

mitosis   The process by which cells divide. In this case mecha-
                 nisms in the mitotic cell cycle ensure that the number
                 of **chromosomes** in each daughter cell is the same as in
                 the parent cell.

mitochondria   Small energy-producing structures—lozenge-like cap-
                 sules called **organelles**—found inside cells. There are
                 several thousand in most cells. The egg (**oocyte**) at
                 ovulation has a particularly large number, but no new
                 mitochondria are produced before the **blastocyst**
                 stage. The **DNA** of all mitochondria in a body,
                 whether male or female, is obtained from the mito-

chondria in the egg, so this **genetic code** is only passed down the maternal line. Mitochondria give rise to other mitochondria by copying their small piece of mitochondrial DNA and passing one copy of the DNA along to each of the two resulting mitochondria. In the great majority of cases, the mitochondria of cloned offspring are all derived from the egg (oocyte), not the donor of the DNA.

mitosis   The process of **cell division** where the **DNA** is replicated and one cell becomes two.

molecular biology   The study of the molecular basis of life, including the biochemistry of molecules such as **DNA** and **RNA**.

monozygotic   Derived from a single (mono) fertilized egg (**zygote**). Monozygotic twins form when one fertilized egg separates into two identical zygotes. This process (**twinning**) can also be performed in the laboratory.

morula   The **embryo** at the stage when it is a ball of closely knit cells, before it becomes a **blastocyst**.

mutation   A change in genes produced by a chance or deliberate change in the **DNA** that makes up the hereditary recipe of an organism.

motor neuron disease   A group of disorders in which motor nerve cells (**neurons**) in the spinal cord and brain stem deteriorate and die. ALS, also called Lou Gehrig's disease after the great baseball player who succumbed, is the most common motor neuron disease.

MPF   Maturation-promoting factor, meiosis promoting factor, mitosis promoting factor, depending on which scientist you talk to. A complex **protein** made up of several protein subunits that acts at a specific point in the cell cycle to cause the breakdown of the envelope that surrounds the nucleus and the packaging of **DNA** into **chromosomes,** so-called chromosome condensation. MPF plays an important role during **nuclear transfer.**

neuron

The nerve cell that is the fundamental unit of the nervous system.

neuro-transmitter

A chemical that transmits signals between nerve cells.

nucleic acid

Complex organic acid made up of a long chain of units called bases. The two types, known as **DNA** and **RNA,** form the basis of heredity.

nuclear transfer

The technical term for the most common form of cloning. Transfer of a nucleus from a donor cell into an egg (**oocyte**) from which the **chromosomes** have been removed. The genetic information in that **nucleus** determines almost all of the characteristics of the resulting offspring.

nucleus

The small region in the center of a cell that contains the **DNA/chromosomes**—the instructions to build and run the cell.

oocyte

This term is used carelessly to describe the female **gamete**—the egg—at several different stages of **meiosis.** At the stage when oocytes are shed from the **ovary,** the oocytes are at the second metaphase of meiosis (MII), which is the preferred stage for use in **nuclear transfer.**

organelles

Structures found within cells that are analogous to organs in a body. These are too small to be seen with a light microscope.

ovary   The female reproductive organ that produces egg cells
                     and the steroid hormones that govern many aspects of
                     reproductive behavior and the function of other
                     organs of the reproductive tract, such as the **uterus**.

ovulation   The release of eggs from the **ovary**.

Parkinson's   A progressive neurodegenerative disorder caused by
disease   damaged or dead **dopamine**-neurons in a region of the
                     brain that controls balance and coordinates muscle
                     movement.

partheno-   The development of **embryos** from unfertilized eggs
genesis   that are activated by stimulation with electric pulses or
                     chemicals. Literally, "virgin birth."

placenta   The organ that is formed during pregnancy to nourish
                     the developing fetus. It is made up of tissue from both
                     the fetus and the mother.

pluripotent   Pluripotent cells are able to give rise to cells of various
                     different types, but not all of the types. Only **totipo-
                     tent** cells can form all cell types.

polar body   The means by which a cell sheds **chromosomes**. The
                     body is a small structure outside the egg (**oocyte**) that
                     contains the chromosomes that were discarded during
                     meiosis. During experiments its presence is used to
                     indicate the stage of meiosis that the oocyte has
                     reached and to provide an indication of the probable
                     location of the chromosomes that have to be removed.

primitive   Thickening in surface of **embryos** that results in the
streak   first clearly recognizable stage in embryonic develop-
                     ment that will go on to form the nervous system. Used

by many scientists as a benchmark to separate an early embryo from a "person."

pronucleus   Small round structure(s) seen within the egg after **fertilization** that contain the **haploid** sets of **chromosomes** (genetic material of egg and sperm) surrounded by a membrane. A normal fertilized egg should contain two pronuclei, one from the egg and one from the sperm.

protein   A large molecule composed of one or more chains of **amino acids** in a specific order, one determined by the gene coding for the **protein**. Proteins are required for the structure, function, and regulation of the body's cells, tissues, and organs. Examples are hormones, **enzymes**, and antibodies.

quiescent   Said of a cell that is comparatively inactive. Some cells in the body are quiescent while they await an instruction from the body to carry out a specific task, but quiescence can be induced in the culture dish either by starving the cells or by allowing them to be overcrowded.

reprogram   A change in the repertoire of **genes** used by a cell. At any particular time the nucleus in a cell is organized to direct the production of a range of **proteins** that is appropriate for its function; a heart cell, for example, would use a spectrum different from that of proteins to a liver cell. The **nucleus** is said to be "reprogrammed" if the organization is changed so that the nature of the cell is changed. The nucleus is also reprogrammed by transfer into an **oocyte**, but this process is inefficient in present procedures, leading to the occurrence of abnormalities of development.

RNA   Short for ribonucleic acid. One of the two types of genetic materials—in the form of nucleic acids—found in all cells. The other is deoxyribonucleic acid (**DNA**). RNA transmits genetic information from DNA to direct the proteins produced by the cell.

RNAi   Short for **RNA** interference. A way to turn off **genes** by introducing double-stranded RNA into a cell.

spindle   Before a normal cell divides, its **chromosomes** are duplicated and then pulled apart by a structure called a **spindle**, so that the two daughter cells each will have the same number of chromosomes. At the end of a normal spindle is the spindle pole, also called the centrosome, which pulls the chromosomes outward.

stem cell   The progenitors of the many different types of cell in the body. In most cases if a cell divides, it is able to give rise only to daughters that are like itself. In addition, such a differentiated cell is able to divide only a limited number of times. By contrast, stem cells have a greater ability to divide and give rise to all other types. Adult stem cells give rise to tissues, and efforts are under way to see how flexible they are.

therapeutic cloning   The production of cells from cloned human **embryos** either for use in research or to treat a disease. The **stem cells** are obtained from the **inner cell mass** of the embryo and then, by the use of growth factors, persuaded to develop into the required type.

totipotency   Totipotent cells are able to give rise to cells of all of the different tissues of an adult. By contrast, **pluripotent** cells give rise only to a limited variety.

twinning Also called blastocyst splitting. The natural form of cloning. Occurs when a **blastocyst** divides in two.

uterus The hollow pear-shaped organ in the lower abdomen of a woman in which the fetus develops. It lies within the pelvis and is also known as the womb.

virus One of the smallest infectious agents, consisting of a piece of **genetic code** wrapped in **protein**, measuring between 15 and 3,000 nanometers across (one nanometer is one-billionth of a meter). They are responsible for a huge range of diseases, such as influenza. It is debatable whether viruses are living, since they have to hijack the molecular machinery of our cells to reproduce (they do this by "reprogramming" our cells with their genetic code, turning them into virus factories). They can themselves be hijacked to carry out **gene therapy**–gene transplants.

xenotrans- The transfer of cells or tissues from one species to
plantation another.

zona pellucida The translucent membrane that encloses the eggs (**oocytes**) and early **embryos** of mammals, such as sheep and humans. The embryo "hatches" from the zona at the **blastocyst** stage of development.

zygote The one-cell **embryo** formed at **fertilization** by the fusion of sperm and egg (**oocyte**).

ACKNOWLEDGMENTS

WE OWE A great debt of gratitude to many people. Very many thanks to John Brockman and Katinka Matson, who provided the encouragement for the two of us to collaborate. Thanks are also due to Angela von der Lippe of Norton in America, and her assistants, Vanessa Levine-Smith and Lydia Winslow Fitzpatrick. For the British edition, we are grateful to Tim Whiting of Little Brown. Neither of us could function in the office without the help of two people: Gulshan Chunara at the *Daily Telegraph* and, at the Roslin and Edinburgh, Lynne Elvin.

The following have answered questions, sent papers, and provided other invaluable advice. Some of the following have also commented on drafts or chapter sections (any remaining howlers are, of course, ours alone). Many thanks to Michael Banner, John Bracken, Derek Bromhall, Ann Bruce, Keith Campbell, Robin Christmas, Alan Colman, Joy Delhanty, Matthew Freeman, Chris Graham, John Gurdon, John Harris, Doris Highfield, Tim King, Robin Lovell-Badge, Dusty Miller, Björn Oback, Jean-Paul Renard, Susan Rhind, Bill Ritchie, Jim Robl, Linda Rosenbaum, Jane Skok, Davor Solter, Miodrag Stojkovic, Andrzej Tarkowski, Ted Tuddenham, Karen Walker, and Steen Willadsen. Photographs were provided by Norrie Russell of the Roslin Institute and Roger Highfield.

Graham Farmelo and Paul Carter provided crucial feedback on how to structure chapters to make the book more accessible. Eamonn Matthews provided transcripts of interviews of his award-winning series on IVF, *Baby Makers*. And Lois Godfrey kindly gave us permission to quote her uncle J. B. S. Haldane.

Finally, heartfelt thanks are due to those closest to us for tolerating our need to invest what little spare time we had in this project. Roger would like to thank his wonderful wife, Julia, for all her loving support, his children, Holly and Rory, for lifting his spirits, and their au pair, Jeanett Svenkerud, and grandmother Betty Brookes for helping to keep them amused. Ian is grateful to his wife, Vivienne, for her continued tolerance of the time that he puts into his work and for her infinite patience and support.

Roger Highfield and Ian Wilmut
January 2006

INDEX

Page numbers in *italics* refer to illustrations.

THE AUTHORS

IAN WILMUT is Professor of Reproductive Science at the Queen's Medical Research Institute for Medical Cell Biology, University of Edinburgh. Until recently he was the head of the Department of Gene Function and Development, Roslin Institute, Edinburgh, Scotland. As leader of the team at the Roslin that in 1996 produced Dolly, the first animal to be cloned from an adult cell, he has played a unique role both in the science of cloning and in the international debate about its implications. Ian has testified before parliamentary and congressional committees in the United Kingdom, France, and the United States and given many public lectures and discussions on the subject.

ROGER HIGHFIELD studied for his doctorate at Oxford University and the Institut Laue-Langevin, Grenoble, and is the award-winning science editor of the *Daily Telegraph* and a member of the UK's Bioscience Futures Forum. Roger has written/coauthored five popular science books, all of which have been translated into foreign editions, including *The Physics of Christmas*, *The Arrow of Time*, and *The Private Lives of Albert Einstein*.